Methods in Molecular Biology

Series Editor
John M. Walker
School of Life Sciences
University of Hertfordshire
Hatfield, Hertfordshire, AL10 9AB, UK

For further volumes:
http://www.springer.com/series/7651

Axon Growth and Regeneration

Methods and Protocols

Edited by

Andrew J. Murray

Department of Biochemistry and Molecular Biophysics, Columbia University, New York, NY, USA

Humana Press

Editor
Andrew J. Murray
Department of Biochemistry
and Molecular Biophysics
Columbia University
New York, NY, USA

ISSN 1064-3745 ISSN 1940-6029 (electronic)
ISBN 978-1-4939-0776-2 ISBN 978-1-4939-0777-9 (eBook)
DOI 10.1007/978-1-4939-0777-9
Springer New York Heidelberg Dordrecht London

Library of Congress Control Number: 2014937663

Printed on acid-free paper

Humana Press is a brand of Springer
Springer is part of Springer Science+Business Media (www.springer.com)

Preface

It is for the science of the future to change, if possible, this harsh decree. Inspired with high ideals, it must work to impede or moderate the gradual decay of the neurons, to overcome the almost invincible rigidity of their connections, and to re-establish normal nerve paths, when disease has severed centres that were intimately associated.

Santiago Ramon y Cajal,
Degeneration and Regeneration of the Nervous System, 1913.

Around 100 years ago, Cajal wrote of how science of the future needed to be able to reestablish normal nerve paths after they had been severed. Today, there has been a great deal of progress in terms of our scientific understanding of how and why the mature mammalian central nervous system (CNS) does not regenerate after injury. However, an effective clinical therapy to repair the damaged CNS still eludes us. Over many years of dedicated and remarkable research, great strides have been made to reveal the physical, cellular, and molecular mechanisms underlying axon growth and why this is prevented in the mature CNS. It now seems clear that this failure is a combination of both an extrinsic inhibitory barrier in the CNS environment and an intrinsic inability of mature neurons to overcome this obstacle.

In this volume, *Axon Growth and Regeneration: Methods and Protocols*, I have aimed to bring together a diverse set of techniques that can be used to study the mechanisms underlying CNS axon growth and consequently, hopefully, provide a resource that will aid in the development of repair strategies. The first part provides a brief perspective outlining some of the current understanding in the field. Parts II and III focus on axon growth in vitro, providing a range of protocols that can be used to examine intracellular signaling pathways, axonal responses to extracellular factors, and methods for quantifying outgrowth. Part IV provides protocols for inducing experimental injury in vivo as well as some highly promising protocols for promoting regeneration. Finally, Part V provides a series of protocols that can be used to monitor the extent of axon regeneration in vivo, ranging from tract tracing to in vivo imaging and functional recovery.

While intended to be extensive, the volume is by no means all inclusive, due to the vastness of the field. However, the range of protocols provided in this volume will hopefully provide a significant set of diverse tools for researchers studying axon regeneration, and I am personally eager to see them used in the next generation of exciting new discoveries in the field. I would like to thank all of the contributors who invested their time and energy in this volume allowing this collection to come to light. I very much hope you enjoy this volume of *Methods in Molecular Biology*.

New York, NY, USA *Andrew J. Murray*

Preface

Contents

Part IV Axon Injury and Regeneration In Vivo

Part V Assessing the Extent of Axon Regeneration

Contributors

HIROKI AKIYAMA • *Laboratory for Neuronal Growth Mechanisms, RIKEN Brain Science Institute, Wako, Saitama, Japan*
RICARDO ALCHINI • *Department of Neurology and Neurosurgery, Montreal Neurological Institute, Montreal, QC, Canada*
FERNANDA MARTINS DE ALMEIDA • *Universidade Federal do Rio de Janeiro, Campus Macaé, Macaé, RJ, Brazil*
JACQUELINE M. BLILEY • *Department of Plastic Surgery, University of Pittsburgh, Pittsburgh, PA, USA*
AHMET BOZKURT • *Department of Plastic Surgery, Reconstructive and Hand Surgery, Burn Centre, Medical Faculty, RWTH Aachen University, Aachen, Germany; JARA - Translational Brain Medicine, RWTH Aachen University, Aachen, Germany*
GARY ANTHONY BROOK • *Medical Faculty, Institute of Neuropathology, RWTH Aachen University, Aachen, Germany; JARA - Translational Brain Medicine, RWTH Aachen University, Aachen, Germany*
HANSRUEDI BUELER • *Department of Microbiology, Immunology, and Molecular Genetics, University of Kentucky College of Medicine, Lexington, KY, USA*
TOM BUYENS • *Neural Circuit Development and Regeneration Research Group, Animal Physiology and Neurobiology Section, Department of Biology, KU Leuven, Leuven, Belgium*
GEOFFREY M.W. COOK • *Department of Physiology, Development and Neuroscience, University of Cambridge, Cambridge, UK*
RONALD DEUMENS • *Institute of Neuroscience, Université Catholique de Louvain (UCL), Brussels, Belgium*
JAMES W. FAWCETT • *John van Geest Centre for Brain Repair, University of Cambridge, Cambridge, UK*
GINO B. FERRARO • *Edwin L. Steele Laboratory of Tumor Biology, Department of Radiation Oncology, Massachusetts General Hospital and Harvard Medical School, Boston, MA, USA*
ALYSON E. FOURNIER • *Department of Neurology and Neurosurgery, Montreal Neurological Institute, Montreal, QC, Canada*
HEIKE FRANKE • *Rudolf Boehm Institute of Pharmacology and Toxicology, University of Leipzig, Leipzig, Germany*
DJOERE GAUBLOMME • *Neural Circuit Development and Regeneration Research Group, Animal Physiology and Neurobiology Section, Department of Biology, KU Leuven, Leuven, Belgium*
LIES DE GROEF • *Neural Circuit Development and Regeneration Research Group, Animal Physiology and Neurobiology Section, Department of Biology, KU Leuven, Leuven, Belgium*
ARUM HAN • *Department of Electrical and Computer Engineering, Texas A&M University, College Station, TX, USA; Department of Biomedical Engineering, Texas A&M University, College Station, TX, USA*

SEUNG BAEK HAN • *Shriners Hospitals Pediatric Research Center, Department of Anatomy and Cell Biology, Temple University School of Medicine, Philadelphia, PA, USA*
CLAUDIA HEINE • *Rudolf Boehm Institute of Pharmacology and Toxicology, University of Leipzig, Leipzig, Germany; Translational Centre for Regenerative Medicine (TRM), University of Leipzig, Leipzig, Germany*
JANOSCH P. HELLER • *John van Geest Centre for Brain Repair, University of Cambridge, Cambridge, UK*
INGE VAN HOVE • *Neural Circuit Development and Regeneration Research Group, Animal Physiology and Neurobiology Section, Department of Biology, KU Leuven, Leuven, Belgium*
THOMAS HAYNES HUTSON • *Neurorestoration Group, Wolfson Centre for Age-Related Diseases, King's College London, London, UK*
PREM JAREONSETTASIN • *Department of Physiology, Development and Neuroscience, University of Cambridge, Cambridge, UK*
HIROYUKI KAMIGUCHI • *Laboratory for Neuronal Growth Mechanisms, RIKEN Brain Science Institute, Wako, Saitama, Japan*
ANDREW KAPLAN • *Department of Neurology and Neurosurgery, Montreal Neurological Institute, Montreal, QC, Canada*
CLAUDIA KATHE • *Neurorestoration Group, Wolfson Centre for Age-Related Diseases, King's College London, London, UK*
ROGER J. KEYNES • *Department of Physiology, Development and Neuroscience, University of Cambridge, Cambridge, UK*
SUNJA KIM • *Department of Veterinary Integrative Biosciences, Texas A&M University, College Station, TX, USA*
HYUKMIN KIM • *Shriners Hospitals Pediatric Research Center, Department of Anatomy and Cell Biology, Temple University School of Medicine, Philadelphia, PA, USA*
JACOB KOFFLER • *Department of Neurosciences, University of California, San Diego, La Jolla, CA, USA*
JESSICA C.F. KWOK • *John van Geest Centre for Brain Repair, University of Cambridge, Cambridge, UK*
JIANRONG LI • *Department of Veterinary Integrative Biosciences, Texas A&M University, College Station, TX, USA*
XUETING LUO • *Department of Neurosurgery, Miami Project to Cure Paralysis, Miller School of Medicine, University of Miami, Miami, FL, USA*
MEREDITH ANN MANIRE • *Shriners Hospitals Pediatric Research Center, Department of Anatomy and Cell Biology, Temple University School of Medicine, Philadelphia, PA, USA*
CLAUDIA MARINANGELI • *Institute of Neuroscience, Université Catholique de Louvain (UCL), Brussels, Belgium*
SUELEN ADRIANI MARQUES • *Laboratório de Regeneração Neural e Função, Departamento de Neurobiologia, Fluminense Federal University, Niterói, RJ, Brazil*
KACEY G. MARRA • *Department of Plastic Surgery, McGowan Institute for Regenerative Medicine, University of Pittsburgh, Pittsburgh, PA, USA; Department of Bioengineering, McGowan Institute for Regenerative Medicine, University of Pittsburgh, Pittsburgh, PA, USA*
ANA MARIA BLANCO MARTINEZ • *Laboratório de Neurodegeneração e Reparo, Centro de Ciências da Saúde, Federal University of Rio de Janeiro, Rio de Janeiro, RJ, Brazil; Instituto de Ciências Biomédicas, Cidade Universitária, Ilha do Fundão, Rio de Janeiro, RJ, Brazil*

SEAN CHRISTOPHER MENEZES • *Neurorestoration Group, Wolfson Centre for Age-Related Diseases, King's College London, London, UK*
LAWRENCE DAVID FALCON MOON • *Neurorestoration Group, Wolfson Centre for Age-Related Diseases, King's College London, London, UK*
LIEVE MOONS • *Neural Circuit Development and Regeneration Research Group, Animal Physiology and Neurobiology Section, Department of Biology, KU Leuven, Leuven, Belgium*
KLAUSS MOSTACADA • *Instituto de Ciências Biomédicas, Centro de Ciências da Saúde, Federal University of Rio de Janeiro, Rio de Janeiro, RJ, Brazil*
ANDREW J. MURRAY • *Department of Biochemistry and Molecular Biophysics, Columbia University, New York, NY, USA*
KEVIN K. PARK • *Department of Neurosurgery, Miami Project to Cure Paralysis, Miller School of Medicine, University of Miami, Miami, FL, USA*
JAEWON PARK • *Department of Electrical and Computer Engineering, Texas A&M University, College Station, TX, USA*
JAMES B. PHILLIPS • *Department of Biomaterials and Tissue Engineering, UCL Eastman Dental Institute, University College London, London, UK*
MARIE-CLAIRE ROONEY • *Neurorestoration Group, Wolfson Centre for Age-Related Diseases, King's College London, London, UK*
EPHRON S. ROSENZWEIG • *Department of Neurosciences, University of California, San Diego, La Jolla, CA, USA*
SAIJILAFU • *Department of Orthopaedic Surgery, The Johns Hopkins University, Baltimore, MD, USA*
RAMSEY F. SAMARA • *Department of Neurosciences, University of California, San Diego, La Jolla, CA, USA*
IOANNA SANDVIG • *MI Lab and Department of Circulation and Medical Imaging, Faculty of Medicine, Norwegian University of Science and Technology, Trondheim, Norway*
AXEL SANDVIG • *MI Lab and Department of Circulation and Medical Imaging, Faculty of Medicine, Norwegian University of Science and Technology, Trondheim, Norway; Department of Neurosurgery, Umeå University Hospital, Umeå, Sweden*
RICARDO SANZ • *Department of Neurology and Neurosurgery, Montreal Neurological Institute, Montreal, QC, Canada*
WESLEY N. SIVAK • *Department of Plastic Surgery, University of Pittsburgh, Pittsburgh, PA, USA*
ANDREW SKUBA • *Shriners Hospitals Pediatric Research Center, Department of Anatomy and Cell Biology, Temple University School of Medicine, Philadelphia, PA, USA*
YOUNG-JIN SON • *Shriners Hospitals Pediatric Research Center, Department of Anatomy and Cell Biology, Temple University School of Medicine, Philadelphia, PA, USA*
STEVEN J. TUCKER • *School of Medical Sciences, University of Aberdeen, Aberdeen, Scotland, UK*
BENJAMIN YUNGHER • *Department of Neurosurgery, Miami Project to Cure Paralysis, Miller School of Medicine, University of Miami, Miami, FL, USA*
BO-YIN ZHANG • *Department of Orthopaedic Surgery, The Johns Hopkins University, Baltimore, MD, USA*
RONG-RONG ZHAO • *John van Geest Centre for Brain Repair, University of Cambridge, Cambridge, UK*
FENG-QUAN ZHOU • *Department of Orthopaedic Surgery, The Johns Hopkins University, Baltimore, MD, USA; The Solomon H. Snyder Department of Neuroscience, The Johns Hopkins University, Baltimore, MD, USA*

Part I

Review

Chapter 1

Axon Regeneration: What Needs to Be Overcome?

Andrew J. Murray

Abstract

Axon regeneration is crucial for recovery of function after nervous system injury. Over many years, research has uncovered numerous factors which prevent damaged axons from regrowing and reforming functional connections after damage. These factors are both extrinsic, relating to the central nervous system environment, and intrinsic, relating to the growth capacity of the neurons themselves. In this short review, I summarize these elements with a view to illustrating how they may be overcome to promote nervous system repair.

Key words Axon regeneration, Axon growth, Growth cone, Spinal cord injury

1 Introduction

> "One having a dislocation in a vertebra of his neck, while he is unconscious of his two legs and his two arms"…"an ailment not to be treated"

A description of a sprain of the vertebrae of the neck from the Edwin Smith Surgical Papyrus, an ancient Egyptian surgical treatise dating from 3000 to 2500BC, describing a case of spinal cord injury and the resultant paraplegia. This description epitomizes the inability of axons in the mature mammalian central nervous system (CNS) to regenerate and reform functional connections after injury, leaving the nervous system unable to control the body. Damage to CNS axons occurs via a wide range of insults, including traumatic injury, inflammation or genetic defects, disrupting communication between the brain and the rest of the body, and leading to severe and persistent functional deficits.

The inability to regrow functional connections is in spite of an attempt from central axons to regrow when damaged. Ramon y Cajal was one of the first to notice that central axons undergo a phase of abortive regenerative attempts after they are damaged, and concluded that an extrinsic factor must be responsible for blocking growth [1]. It was a further 60 years before this extrinsic block was confirmed in seminal studies demonstrating that it was the CNS environment that was likely the cause of regenerative failure.

Andrew J. Murray (ed.), *Axon Growth and Regeneration: Methods and Protocols*, Methods in Molecular Biology, vol. 1162,
DOI 10.1007/978-1-4939-0777-9_1,

Albert Aguayo and colleagues established that injured central axons can regrow when given a permissive environment, in this case a sciatic nerve graft that was transplanted into a spinal lesion site in the mature CNS [2, 3]. These observations led directly to a prevailing hypothesis that it is the environment within the CNS after injury that somehow blocks axonal growth—and engendering promise that the identification, and subsequent blockade, of axonal inhibitors would lead to effective clinical therapies. Great headway has been made in terms of identifying and isolating molecules that act as extracellular inhibitors of axon outgrowth. However, an effective clinical therapy based on neutralization of inhibitory factors still seems relatively out of reach. For this reason, further hypotheses have been developed implicating intrinsic changes within aging neurons that reduce their capacity for growth. This hypothesis is exemplified by demonstrations of the spectacular growth ability of immature neurons when transplanted into the adult CNS [4, 5], seemingly impervious to the "inhibitory" environment around them. These observations, and many others, make it conceivable that a future therapy to promote axon repair will need to take advantage of our knowledge of both extrinsic axon growth factors and intrinsic mechanisms. In this short review, I will first summarize some of the best studied extracellular inhibitors of axon regeneration, before discussing the changes that occur in neurons with age that may sensitize them to their inhibitory effects.

2 Extracellular Inhibitors of Axon Regeneration

In general, extracellular inhibitors of axonal growth can be divided into two broad categories, those associated with myelin (myelin-associated inhibitors; MAIs) and those related to the glial scar, a chemical and physical barrier to axon growth formed after injury (Fig. 1).

2.1 Myelin-Associated Inhibitors

2.1.1 Nogo

In both the CNS and peripheral nervous system (PNS), myelin serves to insulate axons and facilitates action potential propagation [6]. However, early studies in the response of growing axons to CNS and PNS myelin demonstrated that central myelin was particularly inhibitory to axon growth [7]. These studies led to the isolation of an antibody (IN-1) against an inhibitory component of myelin that could neutralize the inhibition in vitro [8]. This same antibody could even be used to promote regeneration of corticospinal tract axons in vivo [9]. Identification of the partial peptide sequence for the bovine antigen to IN-1 led to the identification of Nogo-A, a gene encoding an inhibitory protein in myelin in both rats [10] and humans [11, 12]. Nogo itself exists in three isoforms, A, B, and C, and is a member of the reticulon family (comprising proteins normally associated with the endoplasmic reticulum [13]), but it is Nogo-A which has received the most attention as a MAI.

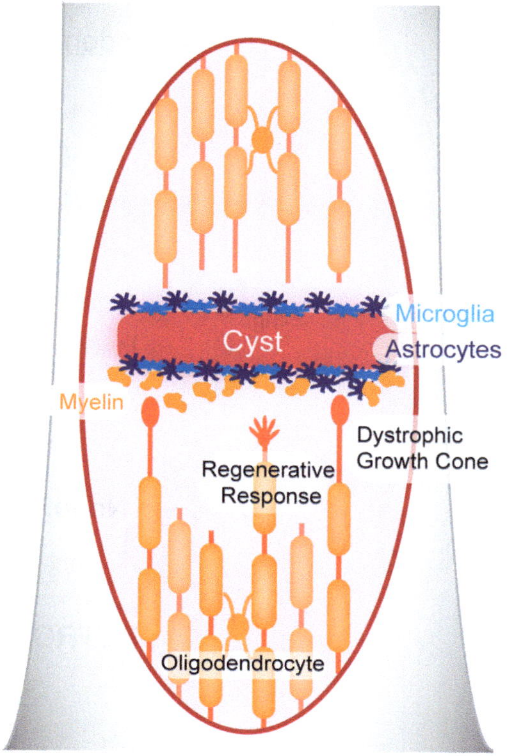

Fig. 1 Summary diagram showing the environment faced by an axon attempting to regenerate. A physical barrier is presented by a cyst as well as infiltrating microglia and reactive astrocytes. Myelin debris and their associated inhibitory molecules also cause growth cones of growing axons to collapse and become dystrophic

Nogo-A has been shown repeatedly to induce growth cone collapse in vitro [10–12]. However, in vivo results using Nogo-A knockout animals have proved difficult to interpret. Several groups have made Nogo-A deletion mutants, alone and in combination with Nogo-B and/or Nogo-C, with variable amounts of regeneration being reported [14–17]. Studies into how Nogo mediates growth inhibition at a cellular level led to the identification of the Nogo-66 receptor (NogoR; [18]), along with a complex of coreceptors that include the p75 neurotrophin receptor ($p75^{NTR}$; [19, 20]), Lingo-1 [21], Taj and TROY [22, 23] (Fig. 2). A C-terminal 66-amino acid loop shared by all Nogo isoforms has allowed the production of NEP1-40 antagonist peptide that has been reported to promote regeneration of the corticospinal tract [24], although perhaps not extensively [25]. More recently, a second receptor for Nogo-A has been discovered, paired immunoglobulin-like receptor B (PirB; [26]). Blocking the activity of PirB, along with the NogoR leads to increased regeneration when compared with either alone [26]. However, use of a PirB antibody in NogoR null neurons only partially blocks nogo-mediated inhibition, perhaps suggesting a further Nogo receptor [27].

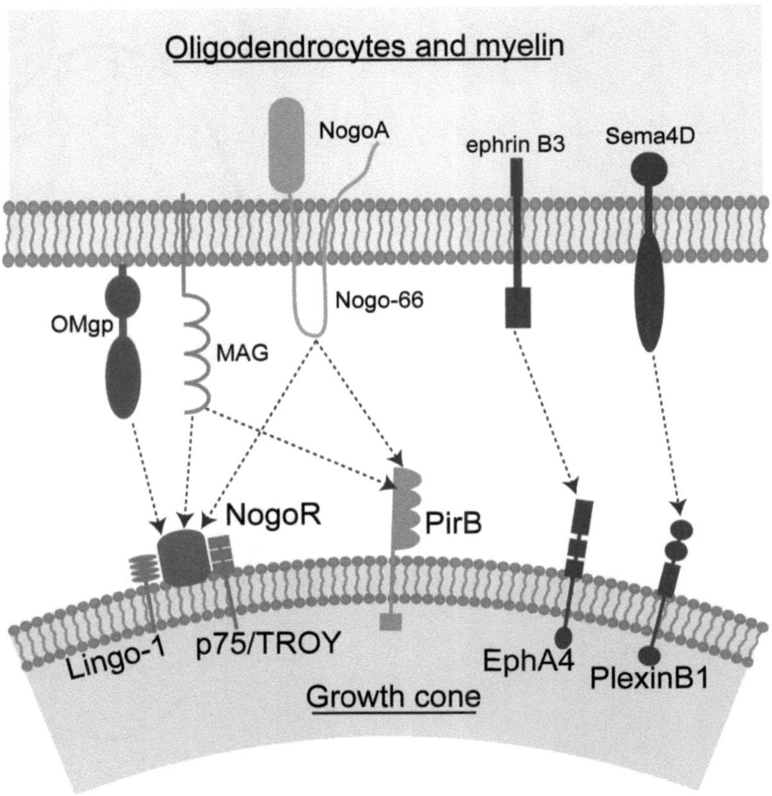

Fig. 2 Summary of inhibitory molecules present on myelin and their associated receptors. Various myelin-associated inhibitors can initiate growth cone collapse by binding to specific receptors on regenerating axons

2.1.2 Myelin-Associated Glycoprotein

Myelin-associated glycoprotein (MAG) was first identified as a component of CNS myelin in the 1970s [28], and is believed to comprise around 1 % of all CNS myelin proteins [29]. In the uninjured nervous system, MAG localizes to the surface of both oligodendrocytes and Schwann cells and functions in the maintenance of myelinated axons [30]. However, upon injury MAG can inhibit growth from a variety of neuronal types [31, 32]; for review *see* [13]. MAG is expressed in both the CNS and PNS; however, MAG is cleared more rapidly in the PNS—likely leading to residual MAG in the CNS which limits axon growth [33]. There is no enhancement of outgrowth by neurons grown on CNS myelin that lack MAG, nor is there enhanced regeneration of the corticospinal tract in MAG null mice [34, 35], perhaps indicating a redundancy between the multiple MAIs present in the CNS.

MAG has been shown to be a high affinity ligand for the NogoR [36, 37], which is surprising given the lack of structural similarity between MAG and NogoA. This interaction continues to require the p75^{NTR} for signal transduction, activating downstream signaling pathways leading to cytoskeletal destabilization [36]. Other receptors mediating MAG growth inhibition include PirB, which can bind MAG with the same affinity as the NogoR [26] (*see* also ref. 38 for review of MAG receptors). Interestingly, MAG is a bifunctional regulator of axon growth, although it is a potent inhibitor to mature neurons both in vitro and in vivo, MAG can actually promote growth from dorsal root ganglion neurons younger than 5 days old [32].

2.1.3 Oligodendrocyte Myelin Glycoprotein

Following the identification of Nogo-A and MAG as inhibitors of neurite outgrowth two independent groups identified oligodendrocyte myelin glycoprotein (OMgp) as a third myelin-associated growth inhibitor. In a screen for glycosylphosphatidylinositol (GPI) linked proteins expressed on CNS myelin one protein with a molecular mass of approximately 110 kDa was found to cause growth cone collapse in embryonic chick DRG neurons. This protein was identified as oligodendrocyte myelin glycoprotein (OMgp; [39]), which had been previously identified on CNS myelin and can be released from myelin by treatment with phosphatidylinositol-specific phospholipase C (PI-PLC; [40]). In a separate study Kottis et al. [41] purified extracts of CNS myelin using agarose chromatography and found that OMgp could inhibit neurite outgrowth from DRG neurons, cerebellar granule cells, and hippocampal neurons.

Interestingly OMgp, along with MAG and Nogo-A, is also a functional ligand of the NogoR [19]. However, generally rodent injury models have shown relatively little regeneration in the absence of OMgp alone [35] again corroborating the hypothesis that there is significant redundancy between the various MAIs.

2.1.4 Other Myelin-Associated Inhibitors: Ephrins, Semaphorins, and Netrins

Several molecular families which act during development to guide growing axons to their targets have been implicated in contributing to the inhibitory environment in the mature CNS. The ephrins and their Eph receptors, for example, are a family of axon guidance molecules that mediate repulsive effects on axons during development [42]. Ephrin-B3 is expressed on mature oligodendrocytes and could be as potent an inhibitor to neurite outgrowth as MAG, Nogo-A, or OMgp [43]. Netrin-1 is a secreted, long-range chemotropic axon guidance cue acting as both a chemoattractant and repellent [44]. In the mature CNS, netrin-1 is selectively enriched adjacent to the paranodal loops of the nodes of Ranvier. Neutralizing netrin-1 in adult rat spinal cord via UNC5 antibodies increases outgrowth from spinal neurons [45]. Further studies are required though to discern the exact contribution Netrin-1 has to the inhibitory CNS environment after injury. Finally, semaphorins generally cause growth cone collapse and axon repulsion during development [46]. In the mature CNS, Sema4D is expressed on oligodendrocytes and is transiently upregulated in the white matter after spinal cord injury [47]. Additionally, Sema5A is expressed on oligodendrocytes of the optic nerve and inhibits neurite outgrowth from retinal ganglion cells [48].

The relative roles that ephrins, semaphorins, and netrin-1 play in the inhibitory CNS environment after injury is not as well studied as the "classical" MAIs, Nogo-A, MAG, and OMgp. Each of these molecules has been extensively studied during development and so many details of the molecular mechanisms underlying their function are known. However, further studies will be required to elucidate their role in the mature CNS and in particular their contribution to a growth inhibitory environment after injury.

3 Glial Scar-Associated Inhibitors of Regeneration

3.1 Physical Components of the Glial Scar

Following physical trauma to the CNS, a number of different cell types mount a response. Initially a cyst may form and macrophages arrive from the bloodstream to remove cellular debris. These are closely followed by oligodendrocyte precursors, recruited from the surrounding tissue and present approximately 3–5 days post injury [49]. The final structural response to injury is astrocytic, with astrocytes proliferating and migrating to the injury site, forming a dense meshwork of cell bodies and processes that constitute a formidable physical barrier to a growing axon (Fig. 1). This "glial scar" matures and finally consists of a tight network of astrocytes bound together by tight and gap junctions [50].

Although the glial scar is a significant impediment to growing axons, it is important to note that it does have an important physiological function. In mice engineered to have little astrocytic response to injury, small traumas, which cause little damage in wild-type animals initiate leukocyte infiltration, neuronal, and oligodendrocyte cell death and pronounced motor defects [51].

3.2 Chemical Components of the Glial Scar

3.2.1 Chondroitin Sulfate Proteoglycans

Once reactive, astrocytes begin to release chondroitin sulfate proteoglycans (CSPGs), a family of extracellular matrix molecules characterized by a core protein and a number of sulfated glycosaminoglycan side chains [52]. Numerous studies, both in vitro and in vivo, have shown that CSPGs actively induce growth cone collapse and block axon growth (*see* refs. 53, 54 for reviews). CSPGs form a strong chemical barrier preventing axon growth, but when combined with the physical barrier of the glial scar this becomes an extremely potent inhibitory combination. Therefore, suppressing the formation of the scar and removal of CSPGs is a major therapeutic target for promoting axon regeneration. One strategy that has emerged, and has showed great promise, is the digestion of the CSPG glycosaminoglycan side chains with a bacterial enzyme ChABC. In vitro ChABC has been shown to enhance neurite outgrowth on CSPG-rich substrates [55, 56]. In vivo local application of ChABC can enhance regeneration of severed axons and increase sprouting by those spared from injury [57, 58]. Even though the inhibitory actions of CSPGs on axon growth has been known for over 20 years, the molecular mechanism of how CSPGs act on growth cones to block growth are still not well understood (for review *see* ref. 52). Several receptors for CSPGs have been reported including PTPσ, LAR phosphatase, NgR1, and NgR3 [59–61]. Additionally, a number of intracellular signaling pathways have been described that impinge CSPG inhibition on to the neuronal cytoskeleton. These have included glycogen synthase kinase 3 [62], protein kinase C [63], and RhoA [64]. Targeting the cellular signaling pathways that initiate growth cone response to CSPGs along with ChABC treatment may well provide a promising strategy for promoting axon growth past the glial scar.

4 Age-Related Decline in Neuron Growth Ability

After injury, axons mount a valiant regenerative response and can construct a growth cone at the severed tip. Upon interaction with MAIs, the glial scar or CSPGs growth cones invariably collapse, changing from a flattened growing protrusion with filopdia and lamelopodia, to a rounded bulb incapable of growth.

While the numerous molecules expressed in the mature CNS make significant axon regeneration a formidable challenge, encouragement can be found by examining neurons in the embryonic CNS. Classic studies [4] showed that embryonic human spinal cord tissue transplanted into a lesioned adult rat spinal cord can project extensive axons, seemingly unperturbed by the inhibitory environment. Furthermore, embryonic neurons in vitro transplanted onto adult rat spinal cord tissue are able to grow extensive processes, whereas postnatal neurons are not [65]. These studies demonstrate that as neurons age they lose a capacity for growth and are more susceptible to growth inhibitors. Interestingly, this decline in regenerative ability with age is not restricted to mammals [66].

5 Molecular Correlates of Age-Related Growth Decline

5.1 Bcl-2

One of the first molecules to be implicated in developmental-dependent reduction of axon growth was the proto oncogene Bcl-2 [67]. Bcl-2 expression is correlated with axon growth during development and overexpression increases the growth capacity of retinal ganglion cells [68]. There is debate, however, regarding whether Bcl-2 directly increases the growth capacity of neuronal axons, or whether it supports cell survival [69, 70].

5.2 Kruppel-Like Factors

Kruppel-like factors (KLFs) are a set of zinc-finger transcription factors which have been thoroughly studied in cancer and play roles in the cell cycle, proliferation, and cell death [71]. In a screen of developmentally regulated genes in retinal ganglion cells, expression of KLFs was found to be altered with age [72]. Genetic knockout of KLF4 leads to enhanced neurite outgrowth in vitro and regeneration of retinal ganglion cells in vivo [72]. Interestingly, other KLFs, such as KLF7, have also been reported to be involved in axon growth [73]. A number of KLFs have been shown to be developmentally regulated, both positively and negatively and can have opposing effects on axon growth. This suggests an important role for this family in neuronal development and presents an interesting target to probe in studies of regeneration (*see* ref. [74] for review).

5.3 Cyclic-Adenosine Monophosphate

Levels of the intracellular second messenger cyclic adenosine monophosphate (cAMP), decline in dorsal root ganglion neurons with age, closely correlating with a loss of ability to grow on MAI expressing substrates [75]. Interestingly, this reduction in cAMP levels during development also occurs in *Xenopus* retinal ganglion cells, initiating a switch from attraction to repulsion to netrin-1 [76]. It has been known for some time that damage to the peripheral branch of dorsal root ganglion neurons can increase the intrinsic growth capacity of these neurons in culture [77]. This can be demonstrated in experimental models by carrying out a preconditioning lesion to the peripheral branch, which promotes regeneration of central fibers after injury [78]. cAMP levels in neuronal cell bodies were one of the first factors to be identified as being upregulated in conditioned ganglia [79]. Furthermore, several groups have demonstrated that artificial stimulation of cAMP levels can promote axon regeneration in vivo [79–81] and overcome growth inhibitory molecules in vitro [82, 83].

These examples merely touch the surface of research on developmentally regulated signaling molecules that are involved in axon guidance (*see* ref. [84] for a more comprehensive review).

6 Summary

The situation facing a damaged axon attempting to regenerate is a complex one. Faced with a myriad of different inhibitory molecules such as those on damaged myelin and the glial scar coupled with an already reduced capacity for growth the task is extremely difficult. It seems that an effective therapy that promotes axon regeneration beyond the lesion site will require interventions that focus both on the inhibitory extracellular environment and in intracellular capacity for growth.

References

1. Ramón y Cajal S (1928) Degeneration and regeneration of the nervous system. Oxford University Press, London
2. Aguayo AJ, David S, Bray GM (1981) Influences of the glial environment on the elongation of axons after injury: transplantation studies in adult rodents. J Exp Biol 95: 231–240
3. David S, Aguayo AJ (1981) Axonal elongation into peripheral nervous system "bridges" after central nervous system injury in adult rats. Science 214(4523):931–933
4. Wictorin K, Bjorklund A (1992) Axon outgrowth from grafts of human embryonic spinal cord in the lesioned adult rat spinal cord. Neuroreport 3(12):1045–1048
5. Wictorin K, Brundin P, Gustavii B, Lindvall O, Bjorklund A (1990) Reformation of long axon pathways in adult rat central nervous system by human forebrain neuroblasts. Nature 347(6293):556–558. doi:10.1038/347556a0
6. Emery B (2010) Regulation of oligodendrocyte differentiation and myelination. Science 330(6005):779–782. doi:10.1126/science.1190927
7. Schwab ME, Thoenen H (1985) Dissociated neurons regenerate into sciatic but not optic nerve explants in culture irrespective of neurotrophic factors. J Neurosci 5(9):2415–2423
8. Caroni P, Schwab ME (1988) Antibody against myelin-associated inhibitor of neurite growth neutralizes nonpermissive substrate properties of CNS white matter. Neuron 1(1): 85–96
9. Schnell L, Schwab ME (1990) Axonal regeneration in the rat spinal cord produced by an antibody against myelin-associated neurite growth inhibitors. Nature 343(6255):269–272. doi:10.1038/343269a0
10. Chen MS, Huber AB, van der Haar ME, Frank M, Schnell L, Spillmann AA, Christ F, Schwab ME (2000) Nogo-A is a myelin-associated neurite outgrowth inhibitor and an antigen for monoclonal antibody IN-1. Nature 403(6768):434–439. doi:10.1038/35000219
11. GrandPre T, Nakamura F, Vartanian T, Strittmatter SM (2000) Identification of the Nogo inhibitor of axon regeneration as a Reticulon protein. Nature 403(6768):439–444. doi:10.1038/35000226
12. Prinjha R, Moore SE, Vinson M, Blake S, Morrow R, Christie G, Michalovich D, Simmons DL, Walsh FS (2000) Inhibitor of neurite outgrowth in humans. Nature 403(6768):383–384. doi:10.1038/35000287
13. Filbin MT (2003) Myelin-associated inhibitors of axonal regeneration in the adult mammalian CNS. Nat Rev Neurosci 4(9):703–713. doi:10.1038/nrn1195
14. Kim JE, Li S, GrandPre T, Qiu D, Strittmatter SM (2003) Axon regeneration in young adult mice lacking Nogo-A/B. Neuron 38(2):187–199
15. Simonen M, Pedersen V, Weinmann O, Schnell L, Buss A, Ledermann B, Christ F, Sansig G, van der Putten H, Schwab ME (2003) Systemic deletion of the myelin-associated outgrowth inhibitor Nogo-A improves regenerative and plastic responses after spinal cord injury. Neuron 38(2):201–211
16. Woolf CJ (2003) No Nogo: now where to go? Neuron 38(2):153–156
17. Zheng B, Ho C, Li S, Keirstead H, Steward O, Tessier-Lavigne M (2003) Lack of enhanced spinal regeneration in Nogo-deficient mice. Neuron 38(2):213–224
18. Fournier AE, GrandPre T, Strittmatter SM (2001) Identification of a receptor mediating Nogo-66 inhibition of axonal regeneration. Nature 409(6818):341–346. doi:10.1038/35053072
19. Wang KC, Kim JA, Sivasankaran R, Segal R, He Z (2002) P75 interacts with the Nogo receptor as a co-receptor for Nogo, MAG and OMgp. Nature 420(6911):74–78. doi:10.1038/nature01176
20. Yamashita T, Higuchi H, Tohyama M (2002) The p75 receptor transduces the signal from myelin-associated glycoprotein to Rho. J Cell Biol 157(4):565–570. doi:10.1083/jcb.200202010
21. Mi S, Lee X, Shao Z, Thill G, Ji B, Relton J, Levesque M, Allaire N, Perrin S, Sands B, Crowell T, Cate RL, McCoy JM, Pepinsky RB (2004) LINGO-1 is a component of the Nogo-66 receptor/p75 signaling complex. Nat Neurosci 7(3):221–228. doi:10.1038/nn1188
22. Mi S, Lee X, Shao Z, Thill G, Ji B, Relton J, Levesque M, Allaire N, Perrin S, Sands B, Crowell T, Cate RL, McCoy JM, Pepinsky RB (2004) LINGO-1 is a component of the Nogo-66 receptor/p75 signaling complex. Nat Neurosci 7(3):221–228. doi:10.1038/nn1188
23. Shao Z, Browning JL, Lee X, Scott ML, Shulga-Morskaya S, Allaire N, Thill G, Levesque M, Sah D, McCoy JM, Murray B, Jung V, Pepinsky RB, Mi S (2005) TAJ/TROY, an orphan TNF receptor family member, binds Nogo-66 receptor 1 and regulates axonal regeneration. Neuron 45(3):353–359. doi:10.1016/j.neuron.2004.12.050

24. Park JB, Yiu G, Kaneko S, Wang J, Chang J, He XL, Garcia KC, He Z (2005) A TNF receptor family member, TROY, is a coreceptor with Nogo receptor in mediating the inhibitory activity of myelin inhibitors. Neuron 45(3):345–351. doi:10.1016/j.neuron.2004.12.040
25. GrandPre T, Li S, Strittmatter SM (2002) Nogo-66 receptor antagonist peptide promotes axonal regeneration. Nature 417(6888): 547–551. doi:10.1038/417547a
26. Steward O, Sharp K, Yee KM, Hofstadter M (2008) A re-assessment of the effects of a Nogo-66 receptor antagonist on regenerative growth of axons and locomotor recovery after spinal cord injury in mice. Exp Neurol 209(2):446–468. doi:10.1016/j.expneurol.2007.12.010
27. Atwal JK, Pinkston-Gosse J, Syken J, Stawicki S, Wu Y, Shatz C, Tessier-Lavigne M (2008) PirB is a functional receptor for myelin inhibitors of axonal regeneration. Science 322(5903):967–970. doi:10.1126/science.1161151
28. Filbin MT (2008) PirB, a second receptor for the myelin inhibitors of axonal regeneration Nogo66, MAG, and OMgp: implications for regeneration in vivo. Neuron 60(5):740–742. doi:10.1016/j.neuron.2008.12.001
29. Quarles RH, Sakuragawa N, Everly JL, Pasnak CF, Webster HD, Trapp BD (1978) A biochemical comparison of Xenopus laevis and mammalian myelin from the central and peripheral nervous systems. J Neurobiol 9(3):217–228. doi:10.1002/neu.480090304
30. Trapp BD (1990) Myelin-associated glycoprotein. Location and potential functions. Ann N Y Acad Sci 605:29–43
31. Quarles RH (2007) Myelin-associated glycoprotein (MAG): past, present and beyond. J Neurochem 100(6):1431–1448. doi:10.1111/j.1471-4159.2006.04319.x
32. McKerracher L, David S, Jackson DL, Kottis V, Dunn RJ, Braun PE (1994) Identification of myelin-associated glycoprotein as a major myelin-derived inhibitor of neurite growth. Neuron 13(4):805–811
33. Mukhopadhyay G, Doherty P, Walsh FS, Crocker PR, Filbin MT (1994) A novel role for myelin-associated glycoprotein as an inhibitor of axonal regeneration. Neuron 13(3):757–767
34. Vargas ME, Barres BA (2007) Why is Wallerian degeneration in the CNS so slow? Annu Rev Neurosci 30:153–179. doi:10.1146/annurev.neuro.30.051606.094354
35. Bartsch U, Bandtlow CE, Schnell L, Bartsch S, Spillmann AA, Rubin BP, Hillenbrand R, Montag D, Schwab ME, Schachner M (1995) Lack of evidence that myelin-associated glycoprotein is a major inhibitor of axonal regeneration in the CNS. Neuron 15(6):1375–1381
36. Cafferty WB, Duffy P, Huebner E, Strittmatter SM (2010) MAG and OMgp synergize with Nogo-A to restrict axonal growth and neurological recovery after spinal cord trauma. J Neurosci 30(20):6825–6837. doi:10.1523/JNEUROSCI.6239-09.2010
37. Domeniconi M, Cao Z, Spencer T, Sivasankaran R, Wang K, Nikulina E, Kimura N, Cai H, Deng K, Gao Y, He Z, Filbin M (2002) Myelin-associated glycoprotein interacts with the Nogo66 receptor to inhibit neurite outgrowth. Neuron 35(2):283–290
38. Liu BP, Fournier A, GrandPre T, Strittmatter SM (2002) Myelin-associated glycoprotein as a functional ligand for the Nogo-66 receptor. Science 297(5584):1190–1193. doi:10.1126/science.1073031
39. Schnaar RL, Lopez PH (2009) Myelin-associated glycoprotein and its axonal receptors. J Neurosci Res 87(15):3267–3276. doi:10.1002/jnr.21992
40. Wang KC, Koprivica V, Kim JA, Sivasankaran R, Guo Y, Neve RL, He Z (2002) Oligodendrocyte-myelin glycoprotein is a Nogo receptor ligand that inhibits neurite outgrowth. Nature 417(6892):941–944. doi:10.1038/nature00867
41. Mikol DD, Stefansson K (1988) A phosphatidylinositol-linked peanut agglutinin-binding glycoprotein in central nervous system myelin and on oligodendrocytes. J Cell Biol 106(4):1273–1279
42. Kottis V, Thibault P, Mikol D, Xiao ZC, Zhang R, Dergham P, Braun PE (2002) Oligodendrocyte-myelin glycoprotein (OMgp) is an inhibitor of neurite outgrowth. J Neurochem 82(6): 1566–1569
43. Goldshmit Y, McLenachan S, Turnley A (2006) Roles of Eph receptors and ephrins in the normal and damaged adult CNS. Brain Res Rev 52(2):327–345. doi:10.1016/j.brainresrev.2006.04.006
44. Cenci MA, Campbell K, Wictorin K, Bjorklund A (1992) Striatal c-fos induction by cocaine or apomorphine occurs preferentially in output neurons projecting to the substantia nigra in the rat. Eur J Neurosci 4(4):376–380
45. Colamarino SA, Tessier-Lavigne M (1995) The axonal chemoattractant netrin-1 is also a chemorepellent for trochlear motor axons. Cell 81(4):621–629
46. Low K, Culbertson M, Bradke F, Tessier-Lavigne M, Tuszynski MH (2008) Netrin-1 is a novel myelin-associated inhibitor to axon growth. J Neurosci 28(5):1099–1108. doi:10.1523/JNEUROSCI.4906-07.2008
47. Castellani V, Rougon G (2002) Control of semaphorin signaling. Curr Opin Neurobiol 12(5):532–541

48. Moreau-Fauvarque C, Kumanogoh A, Camand E, Jaillard C, Barbin G, Boquet I, Love C, Jones EY, Kikutani H, Lubetzki C, Dusart I, Chedotal A (2003) The transmembrane semaphorin Sema4D/CD100, an inhibitor of axonal growth, is expressed on oligodendrocytes and upregulated after CNS lesion. J Neurosci 23(27):9229–9239
49. Goldberg JL, Vargas ME, Wang JT, Mandemakers W, Oster SF, Sretavan DW, Barres BA (2004) An oligodendrocyte lineage-specific semaphorin, Sema5A, inhibits axon growth by retinal ganglion cells. J Neurosci 24(21):4989–4999. doi:10.1523/JNEUROSCI.4390-03.2004
50. Chen ZJ, Negra M, Levine A, Ughrin Y, Levine JM (2002) Oligodendrocyte precursor cells: reactive cells that inhibit axon growth and regeneration. J Neurocytol 31(6–7): 481–495
51. Fawcett JW, Asher RA (1999) The glial scar and central nervous system repair. Brain Res Bull 49(6):377–391
52. Faulkner JR, Herrmann JE, Woo MJ, Tansey KE, Doan NB, Sofroniew MV (2004) Reactive astrocytes protect tissue and preserve function after spinal cord injury. J Neurosci 24(9): 2143–2155. doi:10.1523/JNEUROSCI.3547-03.2004
53. Sharma K, Selzer ME, Li S (2012) Scar-mediated inhibition and CSPG receptors in the CNS. Exp Neurol 237(2):370–378. doi:10.1016/j.expneurol.2012.07.009
54. Properzi F, Asher RA, Fawcett JW (2003) Chondroitin sulphate proteoglycans in the central nervous system: changes and synthesis after injury. Biochem Soc Trans 31(2):335–336, doi: 10.1042/BST0310335
55. Kwok JC, Warren P, Fawcett JW (2012) Chondroitin sulfate: a key molecule in the brain matrix. Int J Biochem Cell Biol 44(4):582–586. doi:10.1016/j.biocel.2012.01.004
56. Nakamae T, Tanaka N, Nakanishi K, Kamei N, Sasaki H, Hamasaki T, Yamada K, Yamamoto R, Mochizuki Y, Ochi M (2009) Chondroitinase ABC promotes corticospinal axon growth in organotypic cocultures. Spinal Cord 47(2):161–165. doi:10.1038/sc.2008.74
57. Busch SA, Horn KP, Silver DJ, Silver J (2009) Overcoming macrophage-mediated axonal dieback following CNS injury. J Neurosci 29(32):9967–9976. doi:10.1523/JNEUROSCI.1151-09.2009
58. Bradbury EJ, Moon LD, Popat RJ, King VR, Bennett GS, Patel PN, Fawcett JW, McMahon SB (2002) Chondroitinase ABC promotes functional recovery after spinal cord injury. Nature 416(6881):636–640. doi:10.1038/416636a
59. Crespo D, Asher RA, Lin R, Rhodes KE, Fawcett JW (2007) How does chondroitinase promote functional recovery in the damaged CNS? Exp Neurol 206(2):159–171. doi:10.1016/j.expneurol.2007.05.001
60. Dickendesher TL, Baldwin KT, Mironova YA, Koriyama Y, Raiker SJ, Askew KL, Wood A, Geoffroy CG, Zheng B, Liepmann CD, Katagiri Y, Benowitz LI, Geller HM, Giger RJ (2012) NgR1 and NgR3 are receptors for chondroitin sulfate proteoglycans. Nat Neurosci 15(5):703–712. doi:10.1038/nn.3070
61. Fisher D, Xing B, Dill J, Li H, Hoang HH, Zhao Z, Yang XL, Bachoo R, Cannon S, Longo FM, Sheng M, Silver J, Li S (2011) Leukocyte common antigen-related phosphatase is a functional receptor for chondroitin sulfate proteoglycan axon growth inhibitors. J Neurosci 31(40):14051–14066. doi:10.1523/JNEUROSCI.1737-11.2011
62. Shen Y, Tenney AP, Busch SA, Horn KP, Cuascut FX, Liu K, He Z, Silver J, Flanagan JG (2009) PTPsigma is a receptor for chondroitin sulfate proteoglycan, an inhibitor of neural regeneration. Science 326(5952):592–596. doi:10.1126/science.1178310
63. Dill J, Wang H, Zhou F, Li S (2008) Inactivation of glycogen synthase kinase 3 promotes axonal growth and recovery in the CNS. J Neurosci 28(36):8914–8928. doi:10.1523/JNEUROSCI.1178-08.2008
64. Powell EM, Mercado ML, Calle-Patino Y, Geller HM (2001) Protein kinase C mediates neurite guidance at an astrocyte boundary. Glia 33(4):288–297
65. Monnier PP, Sierra A, Schwab JM, Henke-Fahle S, Mueller BK (2003) The Rho/ROCK pathway mediates neurite growth-inhibitory activity associated with the chondroitin sulfate proteoglycans of the CNS glial scar. Mol Cell Neurosci 22(3):319–330
66. Shewan D, Berry M, Cohen J (1995) Extensive regeneration in vitro by early embryonic neurons on immature and adult CNS tissue. J Neurosci 15(3 Pt 1):2057–2062
67. Zou Y, Chiu H, Zinovyeva A, Ambros V, Chuang CF, Chang C (2013) Developmental decline in neuronal regeneration by the progressive change of two intrinsic timers. Science 340(6130):372–376. doi:10.1126/science.1231321
68. Chen DF, Schneider GE, Martinou JC, Tonegawa S (1997) Bcl-2 promotes regeneration of severed axons in mammalian CNS. Nature 385(6615):434–439. doi:10.1038/385434a0
69. Cho KS, Yang L, Lu B, Feng Ma H, Huang X, Pekny M, Chen DF (2005) Re-establishing the regenerative potential of central nervous system axons in postnatal mice. J Cell Sci 118(Pt 5):863–872. doi:10.1242/jcs.01658

70. Jiao J, Huang X, Feit-Leithman RA, Neve RL, Snider W, Dartt DA, Chen DF (2005) Bcl-2 enhances Ca(2+) signaling to support the intrinsic regenerative capacity of CNS axons. EMBO J 24(5):1068–1078. doi:10.1038/sj.emboj.7600589

71. Inoue T, Hosokawa M, Morigiwa K, Ohashi Y, Fukuda Y (2002) Bcl-2 overexpression does not enhance in vivo axonal regeneration of retinal ganglion cells after peripheral nerve transplantation in adult mice. J Neurosci 22(11):4468–4477

72. Black AR, Black JD, Azizkhan-Clifford J (2001) Sp1 and kruppel-like factor family of transcription factors in cell growth regulation and cancer. J Cell Physiol 188(2):143–160. doi:10.1002/jcp.1111

73. Moore DL, Blackmore MG, Hu Y, Kaestner KH, Bixby JL, Lemmon VP, Goldberg JL (2009) KLF family members regulate intrinsic axon regeneration ability. Science 326(5950): 298–301. doi:10.1126/science.1175737

74. Laub F, Dragomir C, Ramirez F (2006) Mice without transcription factor KLF7 provide new insight into olfactory bulb development. Brain Res 1103(1):108–113. doi:10.1016/j.brainres.2006.05.065

75. Moore DL, Apara A, Goldberg JL (2011) Kruppel-like transcription factors in the nervous system: novel players in neurite outgrowth and axon regeneration. Mol Cell Neurosci 47(4):233–243. doi:10.1016/j.mcn.2011.05.005

76. Cai D, Qiu J, Cao Z, McAtee M, Bregman BS, Filbin MT (2001) Neuronal cyclic AMP controls the developmental loss in ability of axons to regenerate. J Neurosci 21(13):4731–4739

77. Shewan D, Dwivedy A, Anderson R, Holt CE (2002) Age-related changes underlie switch in netrin-1 responsiveness as growth cones advance along visual pathway. Nat Neurosci 5(10):955–962. doi:10.1038/nn919

78. Richardson PM, Issa VM (1984) Peripheral injury enhances central regeneration of primary sensory neurones. Nature 309(5971):791–793

79. Neumann S, Woolf CJ (1999) Regeneration of dorsal column fibers into and beyond the lesion site following adult spinal cord injury. Neuron 23(1):83–91

80. Qiu J, Cai D, Dai H, McAtee M, Hoffman PN, Bregman BS, Filbin MT (2002) Spinal axon regeneration induced by elevation of cyclic AMP. Neuron 34(6):895–903

81. Qiu J, Cai D, Filbin MT (2002) A role for cAMP in regeneration during development and after injury. Prog Brain Res 137:381–387

82. Neumann S, Bradke F, Tessier-Lavigne M, Basbaum AI (2002) Regeneration of sensory axons within the injured spinal cord induced by intraganglionic cAMP elevation. Neuron 34(6):885–893

83. Murray AJ, Shewan DA (2008) Epac mediates cyclic AMP-dependent axon growth, guidance and regeneration. Mol Cell Neurosci 38(4): 578–588. doi:10.1016/j.mcn.2008.05.006

84. Murray AJ, Tucker SJ, Shewan DA (2009) cAMP-dependent axon guidance is distinctly regulated by Epac and protein kinase A. J Neurosci 29(49):15434–15444. doi:10.1523/JNEUROSCI.3071-09.2009

85. Liu K, Tedeschi A, Park KK, He Z (2011) Neuronal intrinsic mechanisms of axon regeneration. Annu Rev Neurosci 34:131–152. doi:10.1146/annurev-neuro-061010-113723

Part II

Monitoring Intracellular Pathways in Growth Cones

Chapter 2

Analysis of Calcium Signals in Steering Neuronal Growth Cones In Vitro

Hiroki Akiyama and Hiroyuki Kamiguchi

Abstract

Calcium imaging allows us to measure the spatial and temporal changes in intracellular calcium concentration in living cells. Localized calcium elevation often functions as the polarizing signal during guided migration including axon guidance. In this chapter, we describe a protocol to quantitatively monitor the spatiotemporal dynamics of calcium signals in neuronal growth cones in the presence of an extracellular concentration gradient of axon guidance cue.

Key words Growth cone, Axon guidance, Ca^{2+} imaging, Turning assay, Chemotaxis

1 Introduction

Calcium imaging is a widely used technique in laboratories across almost all fields of studies. Indicators for calcium ions (Ca^{2+}) change their fluorescence properties upon Ca^{2+} binding. Therefore, recording the changes in fluorescent intensity of the indicator allows us to quantify changes in intracellular Ca^{2+} concentration ($[Ca^{2+}]$). There are two different types of Ca^{2+} indicators: Chemical indicators and genetically encoded indicators. The first generation of synthetic chemical indicators was UV-excitable and ratiometric indicators. They require either dual-wavelength excitation or dual-emission monitoring [1, 2]. Subsequently, to overcome the disadvantages of UV-excitation, i.e., cytotoxicity and generation of autofluorescence from intracellular constitutions, visible light-excitable indicators have been synthesized [3, 4]. Today there are many commercially available indicators with different excitation and emission spectra that cover a broad range of Ca^{2+} concentrations. As for genetically encoded indicators, modern-day versions are derived from photoproteins isolated from organisms like the luminescent jelly fish [5]. The recent development of genetic engineering technology has produced a number of gene-encoded

Andrew J. Murray (ed.), *Axon Growth and Regeneration: Methods and Protocols*, Methods in Molecular Biology, vol. 1162, DOI 10.1007/978-1-4939-0777-9_2, © Springer Science+Business Media New York 2014

indicators that facilitate the generation of knock-in cell lines and animals, and the selective labeling of specific cell types for in vivo and ex vivo imaging [6, 7]. In dissociated cell culture systems, chemical indicators are, however, better choices because their acetoxymethyl (AM)-conjugated form can easily be loaded into the cytosol and generally provide higher signal to noise ratio. Because growth cones are highly motile, their shape changes during the recording period even if the recording takes only a few minutes. Confounding factors such as changes in thickness can be canceled out by using dual-emission ratiometric imaging. Fura-Red is a dual-excitation indicator. Its peak absorbance shifts from 473 nm to 436 nm upon Ca^{2+} binding and excitation at either wave length results in the emission of red light (longer than 550 nm, [8]). Therefore, as $[Ca^{2+}]$ increases, the intensity of emitted light excited at 473 nm decreases. In combination with another blue light-excitable Ca^{2+} indicator, one can perform dual-emission ratiometric imaging with visible light excitation. There are many reports using this system for monitoring Ca^{2+} dynamics in growth cones [9, 10]. In our laboratory, we employ Fura-Red and Oregon green 488 BAPTA (OGB)-1 for dual-emission ratiometric imaging [11].

The first in vitro chemotaxis assay of neuronal growth cones was reported in 1979, in which a perfusion system was used to generate a stable gradient of nerve growth factor (NGF). In that study, NGF was ejected continuously from a micropipette so that NGF distributes as a gradient along the medium flow [12]. A decade later, Mu-ming Poo's group developed a simpler method for generating a stable gradient of guidance cues using repetitive pulses, rather than the continuous ejection method [13]. Because the duration of a single pulse was short, the distribution of the molecule from each ejecting pulse should ideally be Gaussian, i.e., the distribution of the guidance cue is radial and decays exponentially from the tip of the micropipette. Therefore, a stable gradient can be generated without a perfusion system. This method in combination with optical imaging techniques is now widely used to assess chemotaxis and its associated intracellular signaling pathways [11, 14, 15].

In this chapter, we describe the method for quantitative Ca^{2+} imaging in a growth cone exposed to an NGF gradient.

2 Materials

2.1 Culture Media and Reagents

1. Poly-D-lysine (PDL, Sigma) is dissolved in sterile water and filtrated with 0.45 μm filter to make 10 % (w/v) stock solution. The stock solution, stored at −80 °C, is further diluted to make 0.01 % (w/v) solution. The diluted stock is stored at 4 °C and ready for use.

2. Natural mouse laminin (Invitrogen).
3. Goat anti-human Fc (Jackson) for L1-Fc coating of culture dishes.
4. L1-Fc chimeric proteins that consist of the whole extracellular domain of L1 and the Fc region of human immunoglobulin G is prepared as described [16].
5. RPMI 1640 (Invitrogen) supplemented with 10 % fetal bovine serum (JRH Biosciences) and 20 ng/ml NGF (Promega). Antibiotic-antimycotic, containing penicillin G (100 units/ml), streptomycin (100 μg/ml), and amphotericin B (0.25 μg/ml) (Invitrogen) is also added into the medium.
6. Leibovitz L-15 medium (Invitrogen) supplemented with 20 ng/ml NGF, N2 (Invitrogen), and 750 μg/ml bovine serum albumin (Invitrogen).
7. OGB-1-AM and Fura-Red-AM (Invitrogen) are dissolved in dimethyl sulfoxide (Sigma) to make stock solutions (both 1 mM) and put into an ultrasonic bath for 10 min. Then they are filtrated through 0.22 μm filter (Merck Millipore) by centrifuging at 9,100 × *g*. Store at −20 °C.
8. Cremophor EL (Nacalai Tesque) is dissolved in sterile water to make 10 % stock solution. Store at 4 °C for up to 1 month.
9. Dextran, Alexa Fluor 488 (10,000 MW, Invitrogen) is dissolved in phosphate-buffered saline (PBS, Invitrogen) to make 1 mM stock solution (*see* **Note 1**). Store at −20 °C.

2.2 Animals

Fertilized chicken eggs are obtained from local supplier and incubated at 37.7 °C until embryonic day 9 or 10.

2.3 Apparatus for the Generation of a Microscopic Gradient

1. Borosilicate glass tubings with filament (outer diameter: 1.0 mm, inner diameter: 0.5 mm, length: 10 cm, Sutter Instruments) are pulled using a Flaming/Brown micropipette puller (model P-97; Sutter Instruments) to make micropipettes with 1 μm tip.
2. A motor-drive manipulator (MM-89, Narishige) and a three-axis oil hydraulic micromanipulator (MMO-203, Narishige) for coarse and fine control of the micropipette, respectively. Those manipulators must be fixed rigidly to the microscope via an adaptor.
3. A pneumatic picopump (PV-820, World Precision Instruments) with a holder for micropipettes with 1.0 mm outer diameter (World Precision Instruments).
4. An electric stimulator (SEN-3301, Nihon Kohden) for controlling the picopump.
5. Nitrogen gas cylinder with pressure regulator.

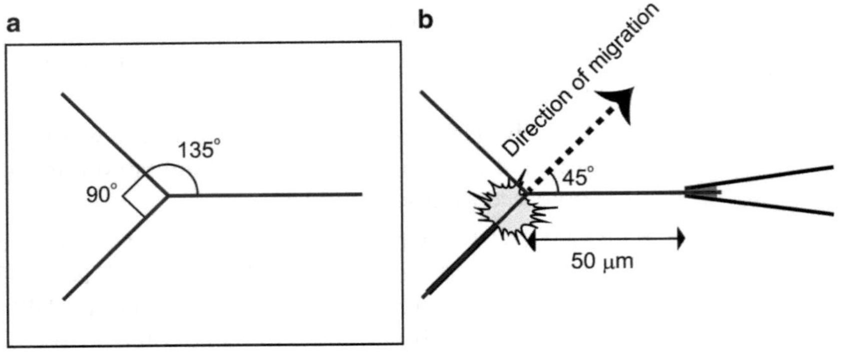

Fig. 1 "Y-sheet" for aligning the micropipette and the growth cone. (**a**) Draw this on a transparent sheet to be attached onto the PC monitor. (**b**) By aligning an axon shaft with either of the two orthogonal lines and positioning the micropipette along the horizontal line, the gradient of the guidance cue of interest can be applied to the growth cone at 45° to its direction of migration

6. The "Y sheet": To facilitate the experiment by showing where the micropipette tip should be positioned with respect to the growth cone (Fig. 1). Affix the "Y sheet" on the monitor of the PC with sticky tape.

2.4 Equipment for Ca^{2+} Imaging

1. An inverted microscope for epifluorescent imaging (IX-81, Olympus) with 20× (UPlanFL N, NA 0.5, Olympus) and 100× (UPlanSApo, NA 1.40, Olympus) objectives. The microscope must be covered by an acrylic box and have a heater in order to maintain the temperature at 37 °C inside the box.
2. A CCD camera (ImagEM, Hamamatsu Photonics).
3. An emission light splitter (Dual View, Roper scientific) with a 565 nm dichroic mirror (DM), and 527/20 nm and 610 nm long pass emission filters (Fig. 2. *See also* **Notes 2** and **3**).
4. Optical filters. In addition to the filters in an emission splitter, an excitation filter (492/18 nm), and a DM (505 nm) are set in the filter wheel with electronic shutter (Ludl Electronic Products) and the microscope, respectively (Fig. 2).
5. A PC with the image acquisition software, MetaMorph (Molecular Devices), that comes with the "split view" function.

2.5 Data Analysis

Software for data quantification, scientific graphing, and statistical analysis: MATLAB 2012a (Math Works) and GraphPad Prism 4.0 (GraphPad Software).

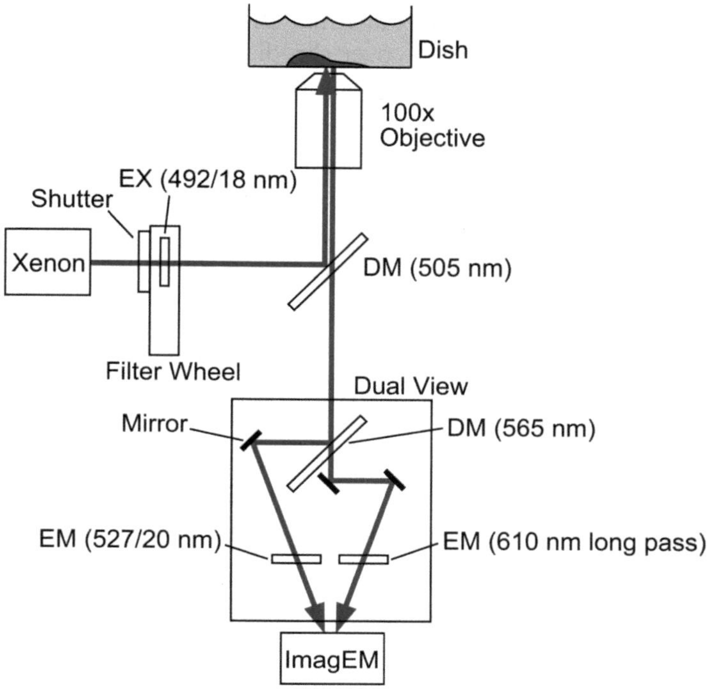

Fig. 2 Schematic drawing of the microscope setup for dual-emission imaging. *EX* excitation filter, *DM* dichroic mirror, *EM* emission filter

3 Methods

3.1 Check for Gradient Generation

1. Switch on the heater inside the acrylic box. Humidify the setup by hanging a moistened paper towel with its bottom tip dipped in a beaker containing distilled water. Wait at least 1 h for the temperature and humidity to stabilize.
2. Mount a glass-based dish containing pre-warmed 2 ml Leibovitz L-15 medium on the microscope.
3. Fill a micropipette with ~2 μl Alexa Fluor 488-dextran solution (1 mM), and connect it to the picopump holder. Then mount the holder onto the micromanipulator.
4. With 20× objective, set the micropipette at 50 μm from the center of the field of view (*see* **Note 4**).
5. Perform imaging as follows: For now, bypass the Dual View unit for full view imaging. Take pictures at 3–5 s intervals with the ImagEM (binning set at 1 × 1) over a time period of 10 min. Within the 10 min, take 30 images as background, then start Alexa ejection (2 Hz, 20 ms duration, 4 psi) (*see* **Note 5**).

3.2 Dish Coating

1. Put 1 ml PDL solution (0.01 %) onto a glass-based dish (bottom cover slip diameter: 27 mm, IWAKI). Incubate at 37 °C for 1 h (up to overnight).
2. Wash the dish twice with sterile water.
3. Put laminin solution (~50 ng/cm^2) on the dish, and incubate at 37 °C for 1 h; or alternatively,
   (a) Put goat anti-human Fc solution (~4 μg/cm^2) on and incubate at 37 °C for 1 h (up to overnight).
   (b) Wash the dish three times with PBS.
   (c) Put 0.5 ml L1-Fc solution on and incubate at 37 °C for 1 h.
4. Rinse with serum-containing RPMI 1640.

3.3 Cell Culture

1. Dissect dosal root ganglia (DRG) from the lumbrosacral region of an embryonic day 9 or 10 chick into an eppendorf tube containing 1 ml Leibovitz L-15 medium.
2. Spin down at 9,100 × *g* for 1 min at 4 °C.
3. Discard supernatant and resuspend in 100 μl trypsin–EDTA (0.05 %, Invitrogen).
4. Incubate at 37 °C for 18 min.
5. Spin down at 9,100 × *g* for 1 min at 4 °C.
6. Discard supernatant and wash in 1 ml Leibovitz L-15 medium.
7. Spin down at 9,100 × *g* for 1 min at 4 °C.
8. Discard supernatant and resuspend in 30 μl Leibovitz L-15 medium.
9. Gently triturate with the 2–20 μl pipette (Eppendorf) without introducing bubbles until a single cell suspension is obtained (~70 strokes usually).
10. Spin down at 9,100 × *g* for 1 min at 4 °C.
11. Discard supernatant and resuspend in 500 μl pre-warmed RPMI1640 medium supplemented with NGF and FBS.
12. Plate cells (~1.4×10^4 cells/dish) onto the glass-based dish coated with either laminin or L1, and incubate in a humidified atmosphere of 95 % air/5 % CO_2 at 37 °C for at least 2 h before loading Ca^{2+} indicators.

3.4 Loading of Ca^{2+} Indicators

All media used here are pre-warmed at 37 °C.

1. Prepare Ca^{2+} indicator solution: Dissolve 2 μl OGB-1 (2 μM final concentration), 2 μl Fura-Red (2 μM final concentration), 0.2 μl Cremophor EL stock solutions in 1 ml Leibovitz L-15 medium. Vortex to mix well.
2. Gently rinse the cell culture with Leibovitz L-15 medium.

3. Incubate cells with the Ca^{2+} indicator solution at 37 °C for 30 min.
4. Gently rinse the culture with Leibovitz L-15 medium.
5. Incubate with 2 ml Leibovitz L-15 medium supplemented with NGF, N2, and BSA at 37 °C for at least 30 min before mounting onto the microscope.

3.5 Ca^{2+} Imaging in the Growth Cone Exposed to an NGF Gradient

Warm up and humidify the environment inside the acrylic box covering the microscope as described above (*see* Subheading 3.1, **step 1**). Wait at least 1 h before starting recordings, allowing the temperature and humidity to stabilize (*see* **Note 6**).

1. Put a drop of immersion oil onto the 100× objective, and mount the culture dish on top.
2. Fill the micropipette with ~2 μl NGF (100 μg/ml), and connect it to the picopump holder controlled by the micromanipulator.
3. After locating the growth cone for imaging, switch the objective from 100× to 20×.
4. Carefully rotate the cell culture dish to adjust the axon angle using the "Y sheet" (Fig. 1, *see also* **Note 7**).
5. Lower the micropipette to the center of the field of view (*see* **Note 8**).
6. Lift the micropipette out of the culture medium in order to minimize dilution due to backpressure.
7. Switch the objective back to 100× (*see* **Note 9**).
8. Lower the micropipette back down to the original position, and then move it to 50 μm from the growth cone (*see* **Note 10**).
9. Perform imaging as follows: Move the Dual View unit back into the emitted light path. Record over a period of 10 min, taking pictures at 3–5-s intervals with the ImagEM (binning set at 4×4). Within the 10 min of recording, the first 30 images are background recording before NGF ejection. Then start the NGF ejection (2 Hz, 20 ms duration, 4 psi).

3.6 Data Analysis

Custom-made programs written with MATLAB are used for all procedures unless otherwise stated.

1. Creating the ratiometric image: Split the recorded image into OGB-1 and Fura-Red channels with the "split view" function of MetaMorph (those two images must be aligned perfectly in X–Y space). After subtracting the background from each image, divide the OGB-1 image by the Fura-Red image. Create a mask for extracting the growth cone region as follows (*see* **Note 11**): Make two new images by applying Gaussian filters with different kurtosis to the Fura-Red image, and

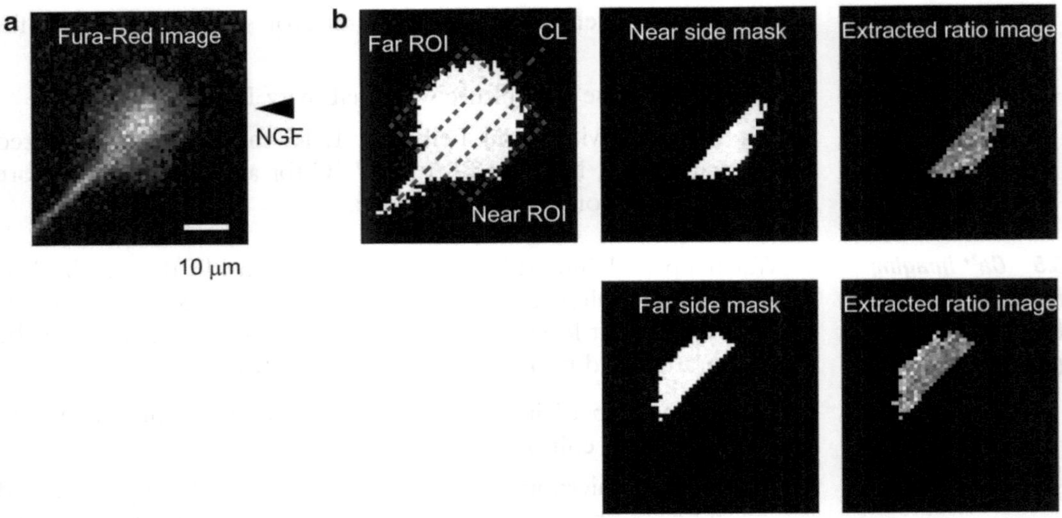

Fig. 3 Extracting the near and far side images of the growth cone for the analysis of asymmetry in Ca^{2+} signals. (**a**) A Fura-Red image of a growth cone loaded with both OGB-1 and Fura-Red. The NGF-containing micropipette was placed on its right side. (**b**) The mask image (*left panel*) is created based on the Fura-Red image (**a**). The center line (CL) and the near and far ROIs are determined as described in the text (*see* Subheading 3.6, **step 2**). By applying the near and far side masks (*middle panels*) over the ratiometric images, the near and far sides of the growth cone are extracted (*right panels*), to be used for further analysis

subtract the more blurred image from the other. Further subtract the resultant image from the original (unprocessed) image. Finally, binarize the last image.

2. Determining the near and far regions of interest (ROIs) (Fig. 3): Draw a 5 μm line along the axon shaft from the point where the shaft connects with the growth cone, start and end points of the line are points "1" and "2". Find the centroid of the growth cone in three consecutive images (points "3," "4," and "5"). A center line for dividing the growth cone into halves is found by taking the linear regression across points 1 to 5. The side facing the NGF source is defined as near side and the opposite as far side. Each growth cone half is further divided into three equal sections along the width (which is perpendicular to the center line). The rectangle enclosing the lateral two-thirds on the near side or far side of the growth cone is defined as the "Near ROI" or "Far ROI," respectively.
3. Calculating the relative changes in $[Ca^{2+}]$ on the near and far sides: R is relative fluorescence between OGB-1 ($F_{OGB\text{-}1}$) and Fura-Red ($F_{Fura\text{-}Red}$).

$$R = \frac{F_{\mathrm{OGB-1}}}{F_{\mathrm{Fura-Red}}}$$

R_0 is the baseline R, taken as the mean of 20 consecutive R before NGF ejection. Calculate R/R_0 at each time point over the period of Ca^{2+} imaging (10 min) on the near and far sides.

4. [Optional] Calculate the near–far asymmetry: Divide the R/R_0 of near side by R/R_0 of far side.
5. Performing statistical analyses with Prism: Calculate the mean R/R_0 during the period from 1 min after the start of NGF ejection till the end of recording for each growth cone. This is because usually the extracellular gradient of NGF takes about a minute to form. Compare R/R_0 on the near side versus the far side using paired t-test to show whether the applied gradient causes any asymmetric changes in $[Ca^{2+}]$. Depending on the experiment, it may be necessary to compare near–far asymmetry under different experimental conditions, for example, in the presence versus absence of a pharmacological inhibitor in the bath medium. In this case, use unpaired t-test or an appropriate multiple comparison test if there are three or more conditions, to assess the effect of drug action on the Ca^{2+} asymmetry.

4 Notes

1. Because the distribution of the ejected molecule theoretically varies depending on the molecular weight, a fluorescent molecule of similar molecular weight as the guidance cue of interest should be used for checking the gradient.
2. OGB-1 and Fura-Red images are simultaneously projected onto the CCD camera, and this image has to be divided into two separate channels to get the ratiometric image for further analysis. Because misalignment causes a critical error in ratiometric imaging (the R value), OGB-1 and Fura-Red images must be aligned carefully by following the alignment protocol provided by the manufacturer.
3. For accurate calculation of the relative changes in $[Ca^{2+}]$, background intensity should be subtracted from the experimental images. To obtain the background intensity, take five images under the same condition as the experimental images, e.g., the filters, illumination intensity, and exposure time, in an empty area of the culture dish.
4. Hints on finding the tip of the micropipette: Reduce the area of the transmitted light as much as possible using the field stop, to generate a very thin column of light. Adjust the position of the micropipette so that the tip falls within the narrowed beam light; lift up the objective, and observe through the eyepiece to find the pipette tip before the tip touches the bottom of the dish; by moving the pipette back and front, you

should be able to find the tip easily; the tip would be most easily found with a phase contrast rather than a differential interference contrast objective; however, even if a differential interference contrast objective is used, raising the condenser will be helpful for finding the tip easily.

5. The distribution is ideally radial. If the distribution skews too much, try: Changing the pressure intensity and/or pulse duration, changing the heater position, changing the position, and/or number of moistened paper towels.
6. The lower limit of humidity for maintaining cells at healthy level is, by our group's experience, 40 %. This can be achieved by directing the heated air flow toward the moistened paper towel.
7. Observe cells through the eyepiece but not through a CCD camera, otherwise you may easily lose the target growth cone when rotating the dish.
8. Remember the *z*-axis location of the micropipette on the micromanipulator so as not to break the tip of the micropipette in later steps.
9. Two types of oils are available for Olympus objectives. Of those two, we use Type-F immersion oil, because it is more viscous than the other. Higher viscosity should minimize spilling of the oil over the 100× objective when switching to the 20× objective.
10. With the 100× objective, Dual View in place, and ImagEM camera, the pipette tip will be outside the field of view before reaching 50 μm apart from the growth cone, thus the distance between the growth cone and the micropipette tip should be measured using the scale on the micromanipulator.
11. Rapid changes in growth cone shape cause confounding factors. Those factors must be eliminated from the analysis, because they cause artifactual increases or decreases in $[Ca^{2+}]$. The factor derived from changes in thickness (*z*-axis) can be canceled out by utilizing ratiometric imaging ($F_{\text{OGB-1}}/F_{\textit{Fura-Red}}$), and the factor due to changes in shape (on the *xy*-plane) can be excluded by extracting the growth cone area from the image with an appropriate mask.

Acknowledgements

This work was partly supported by Grant-in-Aid for Young Scientists (B) 23700456 (H.A.) from the Japan Society for the Promotion of Science. H.A. was supported by the RIKEN Special Postdoctoral Researchers Program. We thank C. Chan and A. T. Guy for their critical reading of the manuscript.

References

1. Tsien RY, Pozzan T, Rink TJ (1982) Calcium homeostasis in intact lymphocytes: cytoplasmic free calcium monitored with a new, intracellularly trapped fluorescent indicator. J Cell Biol 94:325–334
2. Grynkiewicz G, Poenie M, Tsien RY (1985) A new generation of Ca^{2+} indicators with greatly improved fluorescence properties. J Biol Chem 260:3440–3450
3. Minta A, Kao JP, Tsien RY (1989) Fluorescent indicators for cytosolic calcium based on rhodamine and fluorescein chromophores. J Biol Chem 264:8171–8178
4. Takahashi A, Camacho P, Lechleiter JD, Herman B (1999) Measurement of intracellular calcium. Physiol Rev 79:1089–1125
5. Shimomura O, Johnson FH, Saiga Y (1962) Extraction, purification and properties of aequorin, a bioluminescent protein from the luminous hydromedusan, Aequorea. J Cell Comp Physiol 59:223–239
6. Grienberger C, Konnerth A (2012) Imaging calcium in neurons. Neuron 73:862–885
7. Perez Koldenkova V, Nagai T (2013) Genetically encoded Ca indicators: properties and evaluation. Biochim Biophys Acta 1833: 1787–1797
8. Kurebayashi N, Harkins AB, Baylor SM (1993) Use of fura red as an intracellular calcium indicator in frog skeletal muscle fibers. Biophys J 64:1934–1960
9. Gomez TM, Robles E, Poo M, Spitzer NC (2001) Filopodial calcium transients promote substrate-dependent growth cone turning. Science 291:1983–1987
10. Henley JR, Huang KH, Wang D, Poo MM (2004) Calcium mediates bidirectional growth cone turning induced by myelin-associated glycoprotein. Neuron 44:909–916
11. Akiyama H, Matsu-ura T, Mikoshiba K, Kamiguchi H (2009) Control of neuronal growth cone navigation by asymmetric inositol 1,4,5-trisphosphate signals. Sci Signal 2:ra34
12. Gundersen RW, Barrett JN (1979) Neuronal chemotaxis: chick dorsal-root axons turn toward high concentrations of nerve growth factor. Science 206:1079–1080
13. Lohof AM, Quillan M, Dan Y, Poo MM (1992) Asymmetric modulation of cytosolic cAMP activity induces growth cone turning. J Neurosci 12:1253–1261
14. Tojima T, Itofusa R, Kamiguchi H (2010) Asymmetric clathrin-mediated endocytosis drives repulsive growth cone guidance. Neuron 66:370–377
15. Nishiyama M, von Schimmelmann MJ, Togashi K, Findley WM, Hong K (2008) Membrane potential shifts caused by diffusible guidance signals direct growth-cone turning. Nat Neurosci 11:762–771
16. Kamiguchi H, Yoshihara F (2001) The role of endocytic l1 trafficking in polarized adhesion and migration of nerve growth cones. J Neurosci 21:9194–9203

Chapter 3

The Use of Fluorescence Resonance Energy Transfer (FRET) to Measure Axon Growth and Guidance-Related Intracellular Signalling in Live Dorsal Root Ganglia Neuronal Growth Cones

Steven J. Tucker

Abstract

The measurement of signalling by traditional methods in primary neuronal cultures is often limited by cell numbers within the culture and restricted division among these cells. Further limitations are seen with modern fluorescent imaging techniques on account of difficulties with transfection of these cell types. Here, we describe successful transfection of dorsal root ganglion (DRG) primary neuronal cultures with cDNA encoded fluorescence resonance energy transfer (FRET) probes for various signalling moieties, and subsequent measurement of FRET as an index of signalling within these cells. Furthermore, these measurements were made within live neuronal growth cones, which are thin, fragile, and dynamic structures central to axonal growth, repair, and regeneration. This provides novel, physiological insight into the signalling processes driving these axonal behaviors.

Key words Fluorescence resonance energy transfer, Growth cones, Intracellular signalling, DRG neurons, Neuronal signalling, Axon growth, Axon guidance, Live cell imaging

1 Introduction

Axon growth and guidance involves a myriad of signalling interactions involving many soluble factors, cell bound molecules, receptors, and signalling cascades. These interactions occur at the distal tip of the neurite, termed the growth cone, which is responsible for sensing environmental molecules and providing appropriate motile responses [1]. Such responses tend to either be attractive, in which the neurite will grow towards a detected molecule, or repulsive, in which neurite growth is inhibited, the growth cone collapses, or the neurite moves in the opposite direction [2]. Central to many of these growth and guidance "decisions" is cyclic adenosine

Andrew J. Murray (ed.), *Axon Growth and Regeneration: Methods and Protocols*, Methods in Molecular Biology, vol. 1162, DOI 10.1007/978-1-4939-0777-9_3,

monophosphate (cAMP), a cyclic nucleotide second messenger. Indeed, changes in the concentration of cAMP have been proven to influence growth cone turning behaviors in primary cultures of rat dorsal root ganglion (DRG) neurons [3, 4] and in retinal ganglion neurons of *Xenopus laevis* [2]. Therefore, cAMP is a central feature of the growth cones ability to transduce incoming guidance signals.

As an integrative second messenger, cAMP is connected to many signalling cascades and a detailed understanding of these connections remains a key goal in understanding axon guidance. Traditionally, cAMP was thought to be connected only to protein kinase A (PKA), until the recent discovery of a second downstream molecule, exchange protein directly activated by cAMP (Epac) (reviewed in ref. [5]). Through these couplings, cAMP is able to produce a variety of growth cone responses by activating different downstream signalling molecules including B-Raf, extracellular signal regulated kinase (ERK), and protein kinase B (PKB) [4, 6, 7]. In order to delineate this complex signalling network, molecular techniques that permit imaging of fluorescent signalling molecules inside living cells have been applied to DRG neurons. These fluorescent probes utilize fluorescence resonance energy transfer (FRET) to allow visualization and quantitation of signalling processes as they occur in response to guidance molecules.

FRET has recently emerged as an important tool in detecting and quantifying bioactive signalling molecules in living cells [8–11]. Transfection of cDNA encoding fluorescently labelled proteins permits localisation, detection, and measurement of activity within a variety of cellular and biochemical contexts. The underlying theory relates to the fluorophores used to label the proteins of interest. Firstly, 2 fluorophores are used, with overlapping excitation and emission spectra. Specifically, a donor fluorophore with a short excitation wavelength and an emission wavelength that matches the excitation wavelength of the acceptor fluorophore. A commonly used example is pairing cyan fluorescent protein (CFP) (excitation wavelength ~440 nm, emission wavelength ~480 nm) with yellow fluorescent protein (YFP) (excitation wavelength ~480 nm, emission wavelength ~545 nm) [9–12]. Here, excitation of CFP simulates emission of light at a wavelength of ~480 nm. Being close to the excitation wavelength of YFP, this light energy can be captured by the YFP inducing its excitation and subsequent emission of yellow light (~545 nm). It should be noted that other combinations of fluorophores can be utilized, provided they show such overlapping spectra [12]. Secondly, within the structure of the encoded protein of interest, these fluorophores must be held within close molecular proximity (10–100 Å) [12]. By adhering to these design criteria, a system is established that enables the emission wavelength of the donor fluorophore to excite the acceptor fluorophore when the two are in close association.

Measurement of the ratio between donor and acceptor emission permits measurement of protein subunit interaction and subsequent inferences to be made about signalling processes. The efficiency of transfer between the fluorophores is determined by their proximity, as described in the equation below:

$$E = \frac{1}{\left(1 + \left(r - R_0^{\ 6}\right)\right)}$$

Here, E is the transfer efficiency, r is the distance between the fluorophores, and R_0 is the Förster radius (the distance at which the energy transfer is half maximal). Thus, the smaller the distance between the fluorophores, the larger the transfer [12]. The specific design of the FRET probes depends upon the characteristics of the target protein and falls into two categories: intermolecular FRET and intramolecular FRET. The former relates to energy transfer between separate proteins and measures interaction and association of these labelled units. For example, the PKA protein is held quiescent by regulatory subunits, which render its catalytic domains inactive. By labelling these regulatory subunits with the donor fluorophore, and the catalytic subunits with the acceptor fluorophore, maximal FRET is seen when these are bound together and PKA is inactive. Upon activation, these protein subunits dissociate and the transfer of fluorescent energy is lost [9]. Therefore, transfection with 2 cDNA plasmids is required: one encoding the CFP-labelled regulatory subunit and one encoding the YFP-labelled catalytic subunit. In such a system the FRET signal is inversely proportional to the activity of PKA (*see* Fig. 1a).

Intramolecular FRET describes fluorophore labelling at opposing ends of a single protein unit. This "sandwich" conformation is commonly used for proteins that change conformation following activation, e.g., Epac [13]. Hence, when in an inactive, "closed" conformation, the fluorophores are in close molecular proximity and FRET signal is maximal. Upon activation, the "opening" of the Epac protein increases the distance between fluorophores, which can be detected as a steep decrease in energy transfer [13]. Again, FRET signal is inversely proportional to the activity of Epac. This mechanism is represented in Fig. 1b.

Therefore, the precise molecular arrangement of these FRET sensors is context specific and relates to the protein(s) of interest, the interactions involved and their precise molecular development. As this is beyond the scope of this chapter, it is covered within example contexts elsewhere [12, 14].

By utilizing the expression and detection of these fluorescent proteins within the growth cones of DRG neurons, the molecular basis of axon growth, guidance, repair, and regeneration can be studied in live cells.

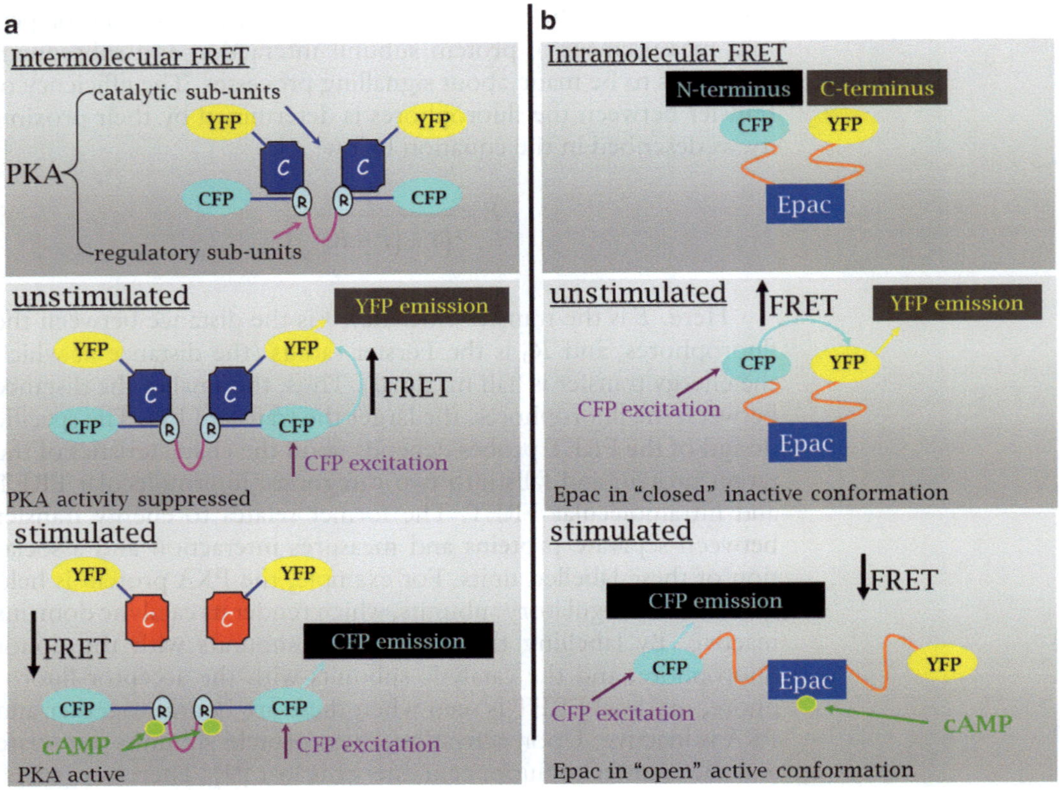

Fig. 1 Intermolecular and intramolecular FRET probes for investigating PKA and Epac signalling. Close molecular proximity between donor (CFP) and acceptor (YFP) fluorophores results in emitted light from the CFP fluorophore (~480 nm) being transferred (by fluorescence resonance energy transfer) to YFP, resulting in emission of light of yellow wavelength from YFP (~545 nm). (**a**) Intermolecular FRET system for investigating PKA signalling. PKA activity is inhibited by association of its catalytic domains with regulatory subunits. By labelling these with acceptor (YFP) and donor (CFP) fluorophores, respectively, FRET signal is maximal when PKA is inactive and decreases upon activation and dissociation of the subunits [9]. (**b**) Intramolecular FRET system for investigating Epac signalling. When inactive, Epac exists in a closed conformation, and opens up upon activation. By labelling the extreme ends of the Epac protein with donor and acceptor fluorophores, these are held closely together and FRET is maximal when the molecule is inactive and decreases upon activation and opening of the Epac structure [13]

2 Materials

2.1 Culture and Transfection of Dorsal Root Ganglion Neurons

Chemicals from Sigma Aldrich unless otherwise stated.

1. Medium for culturing DRG neurons: Bottenstein and Sato's fluid (BSF).
   Prepared as follows: Basal growth medium is Ham's F-12 culture medium (Invitrogen) with the following supplements added: 0.3 % bovine serum albumin (BSA), 16 μg/ml putrescine, 100 U/ml penicillin/streptomycin, 100 μg/ml transferrin, 0.06 μg/ml progesterone, 0.16 μg/ml sodium selenite, 20 μg/ml insulin, and 0.1 μg/ml nerve growth factor (NGF) (Serotec).

2. Culture vessels for FRET measurement: glass bottom culture dishes, 50 mm diameter (MatTek).
3. Appropriate substratum for culturing primary DRG neurons: poly-L-lysine (PLL) and laminin (Invitrogen).
4. Primary culture equipment: 37 °C incubator with 5 % CO_2 supply, Class I or II laminar flow hood.
5. Solution to digest fibrous tissue surrounding ganglia: papain solution.
   Prepared as follows: Papain (10 U/ml), dissolved in retinal buffer (contains 40 mg BSA, 40 mg D–L cysteine, 239 mg HEPES, 578 mg D-glucose, 38 mg calcium chloride, 14 mg magnesium sulfate, 20 mg magnesium chloride, 16 mg sodium bicarbonate, 0.4 ml phenol red solution, and 2 ml 2 M potassium chloride solution made up to 200 ml with Hanks Balanced Salt Solution (HBSS) (Invitrogen)).
6. System for cDNA transfection of DRG neurons: Amaxa Nucleofector machine (Lonza AG), rat neuron Nucleofector solution and kit (Lonza AG).
7. cDNA for transfection of DRG neurons. Must be endotoxin free and of transfection quality.
8. Appropriate pharmacological treatments for manipulating signalling processes of interest.

2.2 FRET Analysis

1. Leica AF6000LX imaging system or equivalent, with appropriate analysis software.

3 Methods

The FRET constructs used to optimize the method described below were designed to elucidate cAMP signalling in DRG neurons and were specifically designed to report on PKA [9], Epac [13], and B-Raf [14]. However, the methodology is broadly applicable to other specific probes, cell types, and systems.

The imaging system used herein incubates samples at 37 °C, but does not have a CO_2 supply, and thus to preserve culture pH throughout the experimental procedures 15 μM HEPES was added as a short-term buffer [15]. To permit application of treatments to the culture during FRET time-lapse experiments, a dish punch was used to create a small hole in the dish lid. This hole must allow pipette tip access to the medium, but not be too large, as this may inhibit cell growth or disrupt cell behavior. Importantly, cells were placed in the incubated microscope unit for 15 min prior to beginning experiments to permit equilibration of the dish, its contents, and the imaging system environment.

3.1 Preparation and Transfection of Neonatal DRG Neurons

The following describes the isolation, transfection, and culture of neonatal rat DRG neurons, and we typically perform such experiments using animals up to 4 days postpartum. This corresponds with a time when these neurons remain capable of regeneration [3–5, 7]. However, such experiments can be readily adapted to other ages and species of animal.

1. Prepare glass bottomed, 50 mm dishes for culturing neurons by adding 500 μl of 10 μg/ml PLL (diluted in sterile, distilled H_2O) to the central well of each dish. Leave dishes within the laminar flow hood overnight.
2. Remove the PLL solution the following morning and dry the dishes thoroughly within the sterile environment of the flow hood (*see* **Note 1**).
3. Once dry, add 500 μl of 5–10 μg/ml laminin solution (diluted in Ham's F-12 medium) to the central well and store in the culture incubator overnight (*see* **Note 2**).
4. In aseptic conditions, carefully dissect the spinal column from a humanely sacrificed animal taking care to remove skin and soft tissues.
5. Place the spinal column in a 60 mm Petri dish containing Ham's F-12 medium and using micro-dissecting scissors, open the midline of the column caudorostrally. Remove the spinal cord, and very carefully reveal the dorsal root ganglia by trimming the lateral walls of the column. The ganglia should become visible within the intervertebral recesses.
6. Using fine, sharp-pointed tweezers, remove each ganglion and place in a fresh dish containing Ham's F-12 medium.
7. Once all the visible ganglia have been collected, trim the roots as close as possible to the body of the ganglion and cut larger ganglia in half (*see* **Note 3**). Once trimmed, transfer ganglia to a sterile 1.5 ml tube containing Ham's F-12 and carefully pellet using a very brief (10–20 s) pulse cycle in a bench-top microcentrifuge (*see* **Note 4**).
8. Gently remove the Ham's F-12 medium from the pelleted ganglia and replace with papain solution, incubating at 37 °C for 20 min.
9. Repeat the centrifugation step above and gently remove the papain solution, replacing with 100 μl Nucleofector solution.
10. Carefully triturate the ganglia using a Gilson P100 pipette tip. Continue this process only for a limited time, until the solution becomes turbid (*see* **Note 5**).
11. Allow the larger pieces of debris to settle in the tube and remove 100 μl of cell suspension, transferring into a Nucleofection electrocuvette (*see* **Note 6**).

12. Add an appropriate amount of cDNA to the electrocuvette and complete the Nucleofection procedure by running it through the Nucleofection machine according to the Manufacturer's recommended protocol, specifically G-13 for rat DRG neurons (*see* **Note 7**).
13. Immediately remove the cells from the cuvette using the micropipette provided in the Nucleofection kit and add to 2 ml fresh BSF (*see* **Note 8**).
14. Use a hemocytometer to count the cells and dilute to a density of 5,000 neurons/ml (*see* **Note 9**).
15. Remove the laminin-coated glass bottomed 50 mm dishes from the incubator and dispose of the laminin solution. Carefully add the transfected neurones onto the central well and replace in the incubator overnight (*see* **Note 10**).
16. Next morning, flood the entire dish with 4 ml of BSF and replace in the 37 °C incubator.
17. Cells will be ready for FRET time-lapse imaging 24–48 h post transfection.

3.2 FRET Set Up, Capture, and Analysis

1. Check cultures for neurite outgrowth and the presence of visible growth cones using a light microscope and supplement medium with 15 μM HEPES for buffering purposes.
2. If pharmacological treatments are to be applied, a lid with a hole punched in should be placed on the cells (*see* **Note 11**).
3. Place the dish of transfected cells in the 37 °C imaging chamber and allow to equilibrate for 15 min.
4. Configure the microscope with appropriate filter sets for the FRET probe design. In total there are three critical channels: the donor channel (donor fluorophore excitation wavelength, donor fluorophore emission wavelength), the FRET channel (donor fluorophore excitation wavelength, acceptor fluorophore emission wavelength), and the acceptor channel (acceptor fluorophore excitation wavelength, acceptor emission wavelength). In addition, a differential interference contrast (DIC) or bright-field channel can be included to help locate the correct focal plane and to focus sharply within it. This should be removed prior to onset of FRET time lapses however, as it will interfere with FRET analysis and signal calculations.
5. Capture sets of control, background, and correction images. These are then used by the FRET software to create analytical correction factors based on an algorithm defined by Wouters et al. [16] (*see* **Note 12**).

6. Locate an appropriate stable growth cone showing fluorescence and check fluorescence in donor, FRET, and acceptor channels (*see* **Note 13**).
7. Run time-lapse experiments by capturing consecutive images using the separate donor, FRET, and acceptor channels (remember to remove the DIC or bright-field channels prior to this). The time between capturing these sets of images should be optimized to minimize bleaching and phototoxicity. In our hands, images taken every 2 min over a total time of 30 min showed fluorescent stability and minimum impact on cellular health (*see* **Note 14**).
8. The three channels are combined to produce a fourth pseudo-color image of the growth cone illustrating FRET sensitized emission (FRET SE), which shows the FRET signal corrected for background and cross talk between the fluorescent channels.
9. Treatments can be applied through the hole in the culture dish lid after four frames have elapsed. This helps to establish a baseline, control FRET SE reading for the specific setup.
10. Analysis is conducted by highlighting regions of interest (ROIs) within the growth cones and reading the FRET SE value from the analysis software. By tracking the ROI throughout the time lapse, changes in FRET SE can be measured before and after treatment or in any other relevant experimental context including asymmetrical responses within the growth cone, monitoring directional signalling gradients, localizing pockets of signalling activity, and quantifying contact-dependent changes in signalling behavior (*see* **Note 15**).

In our hands, this method has proved sensitive, selective, adaptable, and reproducible (*see* Fig. 2) in DRG neurons. Data presented in Fig. 2a show the selective activation of these probes in DRG growth cones using Epac- and PKA-cAMP analogues, suggesting that they are functional and selective in this cellular context. In Fig. 2b the use of these probes to delineate signalling cascades is demonstrated, where netrin-1 specifically activated PKA in neonatal DRG growth cones, but failed to show such activation of Epac. Thus, downstream signalling events can be inferred. Clearly, this modern, novel, and sensitive reporter system has broad potential and is widely applicable across wider contexts within the field of neuroscience.

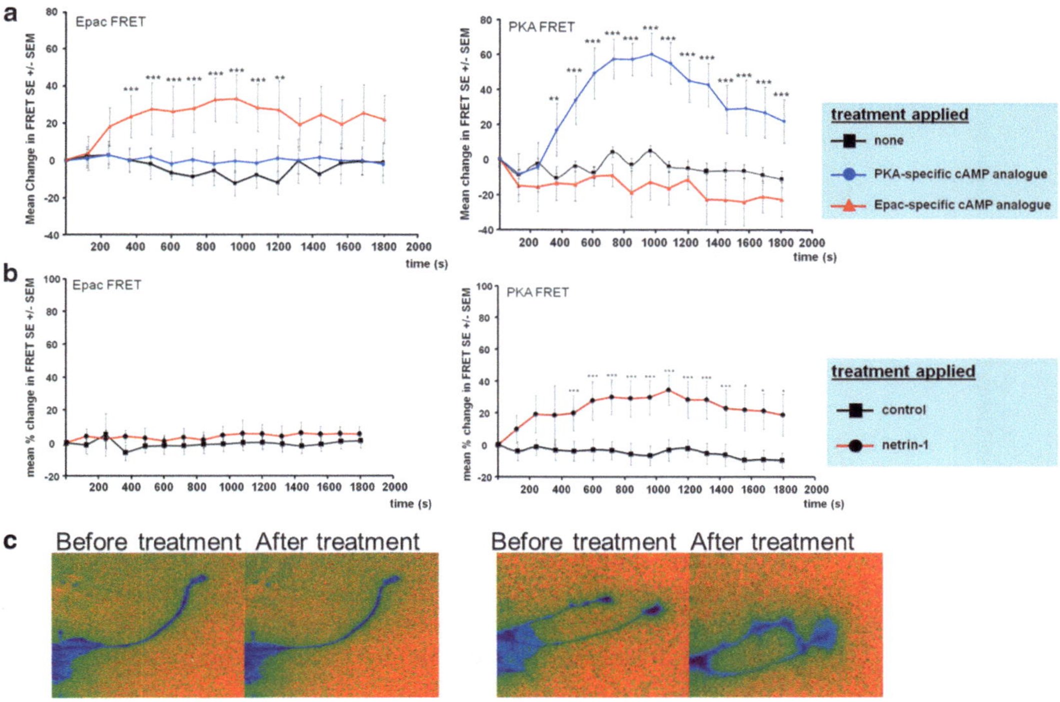

Fig. 2 FRET probes are specific, sensitive, and adaptive for measuring activity in neonatal DRG neuronal growth cones. (**a**) Neonatal (P2) rat DRG neurons were transfected with cDNA encoding the Epac FRET probe (*left panel*) or the PKA FRET system (*right panel*). 48 h later, time-lapse images were taken over 30 min, with frames captured every 2 min. Treatments were applied after 4 min and were Ham's F-12 medium (control), a PKA-specific cAMP analogue (6-Phe-cAMP) (5 μM) (Biolog) or an Epac-specific cAMP analogue (8-pMeOPT-2′-O-Me-cAMP) (2 μM) (Biolog). Results are mean FRET SE ± SEM, $n=7$ for each transfection and treatment combination. Test points were compared to Ham's F12 medium control using Student's unpaired *t*-test; $*p<0.05$; $**p<0.01$; $***p<0.001$. These data indicate the selective activation of these FRET probe systems by specific Epac or PKA analogues. Note, results have been presented as mean % changes in FRET from baseline, and inverted such that positive changes correspond with increased PKA or Epac activity. This makes their interpretation more intuitive. (**b**) As above, but cells were treated with a key axon guidance molecule, netrin-1 (5 ng/ml). Again results are mean FRET SE ± SEM, $n=5$ for each transfection and treatment combination. Test points were compared to Ham's F12 medium control using Student's unpaired *t*-test; $*p<0.05$; $**p<0.01$; $***p<0.001$. This demonstrates specific and selective activation of only the PKA FRET system by netrin-1, a key growth and guidance molecule. Epac remains unaltered. (**c**) Corresponding pseudocolor FRET SE images selected from experiments carried out in (**b**). These show little or no change with time in the FRET signal within growth cones transfected with the Epac probe and treated with netrin-1 (*left hand images*), but dramatic changes within those growth cones transfected with the PKA probe system following application of netrin-1 (*right hand images*). Frames shown are before netrin-1 treatment (240 s) and after netrin-1 treatment (600 s). Note the Epac probe transfected growth cone remains a *dark blue color*, whereas the PKA system transfected growth cone becomes a much *lighter blue color* indicating changes in the FRET signal. Again, these have been inverted so increased FRET indicates increased PKA activation for more intuitive analysis and comprehension

4 Notes

1. Ensure PLL-coated plates are completely dry and free of residual PLL solution, as these will interfere with subsequent substratum application and will influence growth patterns of neurons.
2. Alternative substrata, e.g., fibronectin, collagen can be used dependent upon precise requirements of experiment.
3. The more closely the roots are removed, the fewer Schwann cells will contaminate the final cell number. Care during this procedure lends itself to a more uniform population of actual DRG neurons.
4. When pelleting ganglia, only a very brief centrifugation pulse should be applied to prevent damage to the ganglia and the neurons that they contain.
5. Care should be taking during trituration, and this process should only continue until the solution becomes slightly cloudy. Over-triturating is a key error in the procedure that can drastically reduce viability of the resulting culture.
6. When transferring cells into the electrocuvette, take care to avoid introducing bubbles into the system as these undermine the electroporation procedure.
7. Optimal cDNA quantities need to be determined for individual plasmids using exploratory DNA titration experiments. In our hands, quantity per transfection varies between 2 and 8 μg in neonatal neurons. Interestingly, similar quantities were used for transfection of rat embryonic neurons and higher quantities (10–20 μg per transfection) were required for adult rat neurons.
8. It is vital to minimize the amount of time the cells remain suspended in Nucleofection reagent (particularly true in late neonatal and adult DRG cultures). Therefore, transfer the transfected neurons into BSF immediately after Nucleofection has been completed.
9. The precise density of neurons plated depends upon the needs of the experiment, but 5,000/ml is generally optimal for neonatal rat DRG neurons. Adult and embryonic DRG preparations require lower seeding density on account of their size and growth potential, respectively.
10. Ensure that the laminin substratum does not dry out on glass bottomed 50 mm dishes prior to the addition of diluted neurons, as this will compromise the quality of the culture surface.
11. Successfully treating cells through the hole in the lid is a very sensitive operation if the cell is to remain still and in focus. Significant practice and optimization is recommended to

permit the experimenter to routinely carry this out between frames in the time-lapse experiments.

12. Instruction on the specific sets of control, cross talk, and background images that are required will be given through the analysis software. Each type of software will differ slightly, but all should adhere to the same principles [16].
13. An ideal growth cone should be well spread out with visible protrusions (filopodia and lamellipodia). There should be a suitable level of fluorescence, although this is exposure/intensity adjustable (*see* Fig. 2).
14. Throughout the setup procedure care should be taken to avoid excessive and unnecessary exposure of sample cells to fluorescent or bright-field light. Minimizing bleaching in this way will preserve the life of the culture. Additionally, bleaching control experiments should be conducted with each FRET probe used to optimize the duration of time-lapse experiments and to determine ideal interval between frames.
15. In the case of FRET probes where decreased fluorescence transfer indicates increased signalling, it is possible to invert calculation and analysis such that increased signalling is represented by increased FRET SE signal. This is done through the software and makes the data and analysis more intuitive.

References

1. Song H, Poo MM (1999) Signal transduction underlying growth cone responses to diffusible factors. Curr Opin Neurobiol 9:355–363
2. Filbin MT (2003) Myelin-associated inhibitors of axonal regeneration in the adult mammalian CNS. Nat Rev Neurosci 4:703–713
3. Shewan D, Dwivedy A, Anderson R, Holt CE (2002) Age related changes underlie switch in netrin-1 responsiveness as growth cones advance along the visual pathway. Nat Neurosci 5:955–962
4. Murray AJ, Tucker SJ, Shewan DA (2009) cAMP-dependent axon guidance is distinctly regulated by Epac and protein kinase A. J Neurosci 29:15434–15444
5. Peace AG, Shewan DA (2011) New perspectives in cyclic AMP-mediated axon growth and guidance: the emerging epoch of Epac. Brain Res Bull 84:280–288
6. Bos JL (2003) Epac: a new cAMP target and new avenues in cAMP research. Nat Rev Mol Cell Biol 4:733–738
7. Murray AJ, Shewan DA (2008) Epac mediates cAMP-dependent axon growth, guidance and regeneration. Mol Cell Neurosci 38:578–588
8. Nagai Y, Miyazaki M, Aoki R, Zama T, Inouye S, Hirose K, Iino M, Hagiwara M (2000) A fluorescent indicator for visualising cAMP-induced phosphorylation in vivo. Nat Biotechnol 18:313–316
9. Zaccolo M, Pozzan T (2002) Discrete microdomains with high concentration of cAMP in stimulated rat neonatal cardiac myocytes. Science 295:1711–1715
10. DiPilato LM, Cheng X, Zhang J (2004) Fluorescent indicators of cAMP and Epac activation reveal differential dynamics of cAMP signalling within discrete subcellular compartments. Proc Natl Acad Sci U S A 101: 16513–16518
11. Lohse MJ, Bünemann M, Hoffman C, Vilardaga JP, Nikolaev VO (2007) Monitoring receptor signalling by intramolecular FRET. Curr Opin Pharmacol 7:547–553
12. van der Krogt GNM, Ogink J, Ponsioen B, Jalink K (2008) A comparison of donor-acceptor pairs for genetically encoded FRET sensors: application to the Epac cAMP sensor as an example. PLoS One 3:e1916. doi:10.1371/journal.pone.0001916

13. Ponsioen B, Zhao J, Riedl J, Zwartkruis F, van der Krogt G, Zaccolo M, Moolenaar WH, Bos JL, Jalink K (2004) Detecting cAMP-induced Epac activation by fluorescence resonance energy transfer: Epac as a novel cAMP indicator. EMBO Rep 5: 1176–1180
14. Terai K, Matsuda M (2006) The amino-terminal B-Raf-specific region mediates calcium-dependent homo- and hetero-dimerization of Raf. EMBO J 25:3556–3564
15. Shipman C Jr (1969) Evaluation of 4-(2-hydroxyethyl)-1-piperazineethanesulfonic acid (HEPES) as a tissue culture buffer. Proc Soc Exp Biol Med 130:305–310
16. Wouters FS, Verveer PJ, Bastiaens PIH (2001) Imaging biochemistry inside cells. Trends Cell Biol 11:203–211

Part III

Assessing Axonal Responses to Extracellular Factors In Vitro

Chapter 4

Neurite Outgrowth and Growth Cone Collapse Assays to Assess Neuronal Responses to Extracellular Cues

Andrew Kaplan, Ricardo Sanz, Gino B. Ferraro, Ricardo Alchini, and Alyson E. Fournier

Abstract

The identification of molecular processes involved in regulating neurite outgrowth is an active area of interest for investigators studying neural development and regeneration. In vitro assays designed to measure growth cone morphology and neurite length are frequently used to assess neuronal responses to developmental guidance cues and inhibitory cues that exist in the adult CNS. Here, we describe the procedures to assess morphological responses of cultured dorsal root ganglion neurons to attractive and repellent cues, with a focus on repellents found in the injured adult CNS. The chapter describes methods to culture the DRGs, apply inhibitory ligands, and assess morphological responses. These assays provide biological readouts to assess the capacity of a molecule to act as an inhibitory or growth promoting cue. The readouts can be used as screening tools to aid in the identification of novel targets or drugs for promoting nerve regeneration.

Key words Primary neuronal cultures, Axon, Regeneration, Growth cone, Myelin, Dorsal root ganglia, Microdissection, Neurite length, Neural development

1 Introduction

The development of neurons into polarized structures with axons and dendrites can be captured in vitro [1]. Neurite extension is a complex process that requires precise orchestration of the intracellular cytoskeleton in response to a wide array of extracellular cues, which can be soluble, membrane-associated, or present in the extracellular matrix. These cues regulate axon outgrowth by acting on receptors located at dynamic sensory tips, called growth cones, of pathfinding axons. In the adult central nervous system (CNS), injured axons have a limited capacity to regenerate. Transected axons may reestablish growth cones in an effort to project to targets, but myelin-associated inhibitors (MAIs) including myelin-associated

Andrew Kaplan, Ricardo Sanz, Gino B. Ferraro, and Ricardo Alchini contributed equally to the manuscript.

Andrew J. Murray (ed.), *Axon Growth and Regeneration: Methods and Protocols*, Methods in Molecular Biology, vol. 1162,
DOI 10.1007/978-1-4939-0777-9_4, © Springer Science+Business Media New York 2014

glycoproteins (MAG), Nogo, and oligodendrocyte myelin glycoprotein (OMgp) as well as glial scar components such as chondroitin sulfate proteoglycans (CSPGs) actively collapse growth cones and contribute to failed regeneration [2]. Extracellular inhibitors, together with a failure to reinitiate intrinsic growth programs, result in a lack of functional recovery following traumatic injury and pathological insults such as stroke, neuroinflammation, and neurodegeneration.

The identification of molecular processes involved in regulating axon regeneration is an active area of investigation. In vitro assays designed to measure the effects of repellent cues on growth cone collapse and neurite extension are important tools to identify molecules with repellent activity, to characterize their molecular mechanism of action, and to screen for reagents that may overcome the influence of inhibitory cues [3–6]. Neurites are severed and retract during the dissection and dissociation of neural tissue and the process of neuritogenesis and neurite extension can then be observed in primary culture of dissociated neurons or tissue explants. Gene expression in neuronal cultures can be readily manipulated through viral transduction or electroporation and intracellular signaling can be targeted through drug delivery. By simultaneously exposing cultured neurons to substrates or soluble ligands that are known to be present at sites of injury in the CNS, important insight into the effects of such manipulations on regenerative capacity can be gained by analyzing neurite length and growth cone collapse. These measures thus serve as valuable biological readouts and can elucidate molecular mechanisms involved in axon regeneration. Various types of mammalian CNS neurons have been used to study responses to repellent cues. Here, we focus on chick dorsal root ganglion (DRG) neurons, which are an attractive model because they possess large growth cones, extend neurites rapidly, and are sensitive to adult mammalian CNS repellents. We include protocol modifications for analysis of rodent DRG neurons. Specifically, we describe procedures to dissect and culture chick DRG neurons, apply candidate inhibitory cues, fix and stain cultures, and assess growth cone morphology and neurite outgrowth. We briefly discuss viral-mediated gene delivery as an approach to manipulate gene expression prior to the application of inhibitory cues.

2 Materials

2.1 Animals and Dissection (See Note 1 for Rodent Modifications)

1. *Fertilized chicken eggs*: For precise staging of chicken embryos, incubate fertilized eggs, with the air cell facing up, in a 37 °C humidified incubator with rocking shelves. Stage E12–E16 embryos are appropriate to assess responses to repellents in the adult CNS, including MAIs and CSPGs, while E6–E8 chick DRGs are generally used to assess responsiveness to developmental guidance cues [7].

2. Dissection tools.
   (a) Stereo-binocular microscope.
   (b) Two pairs of Dumont #5 forceps (Fine Science Tools).
   (c) Disposable transfer pipets.
   (d) 2.5 mm cutting edge spring scissors.
   (e) Petri dishes.
   (f) 5 mL pasteur pipets.
   (g) 1,000 μL pipet and pipet tips.
3. Solutions for DRG dissection.
   (a) Leibovitz's L-15 medium.
4. *Solutions for DRG neuron dissociation* (*see* **Note 2** for rodent modifications).
   (a) 0.25 % Trypsin-EDTA.
   (b) Heat-inactivated fetal bovine serum (FBS).

2.2 Cell Culture

1. *Growth medium* (*see* **Note 3** for rodent modifications).
   (a) Ham's F12 nutrient mixture.
   (b) B27 supplement.
   (c) Penicillin/streptomycin.
   (d) Nerve Growth Factor (NGF).

   Prepare growth medium consisting of F12 medium supplemented with 2 % B27 and 10 ng/mL NGF. NGF is essential for DRG neuron survival. Penicillin and streptomycin may be added as per manufacturer's instructions to avoid bacterial contamination. Equilibrate the growth medium at 37 °C, 5 % CO_2 prior to use.
2. Substrates.
   (a) Poly-L-lysine: Dissolve poly-L-lysine in distilled water to a final concentration of 100 μg/mL (equivalent to 0.1 % w/v).
   (b) Laminin (Engelbreth-holm-swarm mouse tumor-derived; commercially available): Dilute laminin in phosphate-buffered saline (PBS) to a final concentration of 10 μg/mL.
   (c) Candidate growth promoting or inhibitory cues, e.g., myelin membrane [8], recombinant Nogo, or CSPGs (commercially available).

2.3 Fixation and Immunofluorescence

1. *Fixative*: 4 % Paraformaldehyde (PFA) in PBS containing 20 % sucrose.
   (a) Heat 1× PBS (pH 7.4) containing 20 % sucrose to 55 °C.
   (b) Add PFA to final concentration of 4 % w/v and swirl.
   (c) Add approximately ½ drop of 1 N NaOH per 1 g of PFA.
   (d) Continue swirling at 55 °C until PFA dissolves.

(e) pH to 7.4 with HCl or NaOH. Verify pH using a litmus paper strips.

(f) Filter through Whatman blotting paper.

2. Immunofluorescence

(a) Neuron-specific beta-III tubulin MAb (clone TuJ-1) (1:2,000 dilution in PBS containing 1 % BSA).

(b) Fluorescent-conjugated anti-mouse secondary antibody.

(c) Nuclear stain (e.g., Hoechst and DAPI) (1:2,000 dilution in PBS containing 1 % BSA).

(d) Rhodamine-phalloidin (1:1,000 dilution in PBS containing 1 % BSA).

3 Methods

3.1 Substrate Preparation

Substrates should be prepared prior to the dissection. Basal neurite outgrowth from both explant and dissociated cultures is promoted when culture dishes are coated with poly-L-lysine and laminin.

1. Coat tissue culture plates with poly-L-lysine solution for 1 h at room temperature.
2. Remove poly-L-lysine and wash three times with sterile PBS. Excess soluble poly-L-lysine can be toxic, resulting in poor survival.
3. Next, coat the tissue culture plates with laminin for 1 h at 37 °C. Remove the laminin and wash three times with PBS before plating DRG neurons. Do not allow the laminin coating to dry, as this will reduce neuronal adhesion and growth. PBS may be left in the culture dish at room temperature during the dissection. Long-term storage of pre-coated dishes is not advised, as this will reduce laminin bioactivity.

3.1.1 Inhibitory Substrates and Treatments

A range of substances that impair CNS regeneration may be used as culture substrates to study the capacity of neurons to extend neurites on inhibitory substrates. MAIs and CSPGs are widely studied inhibitory ligands [9]. By definition, MAIs are components of intact myelin or myelin debris, which inhibit axon regeneration. CSPGs are produced by reactive astrocytes that form the glial scar and are decorated with highly sulfated glycosaminoglycan (GAG) chains [10, 11]. There is evidence that CSPGs inhibit axon regeneration through both nonspecific charge interference and through inhibitory signaling of the CSPG protein core through axonal receptors [12].

The mode of application of inhibitory cues differs for growth cone collapse and neurite outgrowth assays. The growth cone collapse assay requires that neurons project expanded growth cones on a permissive substrate prior to acute treatment with a soluble inhibitor. For neurite outgrowth assays, dissociated neurons can be plated on coated surfaces or exposed to soluble inhibitory cues over an extended period (For examples *see* refs. [13, 14]). Below, we describe the procedures for preparing treatments with various inhibitory molecules for both substrate and soluble applications. Dose curves should be performed for each treatment to determine optimal treatment concentrations.

Inhibitory Cues as a Substrate

Inhibitory cues can be applied to the culture dish surface through drying or coating. We will describe the common protocol for most inhibitors and then give specific considerations for some commonly used inhibitory molecules.

*Drying Method: Preferable for Myelin Membranes (See **Note 5** for Consideration of Appropriate Controls)*

1. Pipet 50 μL of the diluted inhibitor (in PBS) into individual wells of a 96-well plate and leave to dry overnight in the tissue culture hood, with the cover ajar.
2. The next day, wash the wells with PBS and then coat each well with poly-L-lysine for 1 h. Exceeding the coating time may interfere with the inhibitory effect of the substrate.
3. Remove the poly-L-lysine and wash three times with PBS.
4. Coat the plate with 1 μg/mL laminin for 1 h at 37 °C.
5. Remove the laminin and wash three times 5 min with PBS.

Coating Method: Myelin Membranes

An alternative to drying down myelin is to coat it onto a nitrocellulose layer.

1. Prepare 1 % nitrocellulose: Dissolve 0.1 g Hybond ECl Nitrocellulose in 10 mL methanol. Shake vigorously for 1 h and then centrifuge for 10 min to pellet any impurities. Decant to a new tube.
2. Add 25–50 μL per well in a 96-well plate to make a thin layer of nitrocellulose covering the surface.
3. Let the solution dry for 30 min or until the solution solidifies.
4. Block 1 h at room temperature with 1 % BSA diluted in PBS (0.2 μm filtered just before use).
5. Add diluted inhibitor (in PBS) onto the nitrocellulose layer and incubate for 2 h at room temperature.
6. Gently rinse two times with PBS. Be advised that the nitrocellulose layer can easily peel off if the washes are too vigorous.
7. Coat the plate with 1 μg/mL laminin for 1 h at room temperature. Remove the laminin and gently wash three times 5 min with PBS. Leave PBS in the wells during the dissection to prevent the substrates from drying.

Coating Method: Purified Recombinant Proteins

In our hands, recombinant proteins such as individual MAIs or CSPGs such as aggrecan demonstrate reliable inhibitory activity when coated.

(a) Dilute the stock of recombinant protein in ice-cold PBS.

(b) Coat a poly-L-lysine-coated 96-well plate with diluted MAI or CSPG and 10 μg/mL laminin and mix for 2 h at 37 °C.

(c) Remove MAI or CSPG and wash three times 5 min with PBS.

Substrate specifications

Myelin: Myelin can be prepared from fresh rodent or bovine white matter, with bovine brain offering the advantage of higher yield. Detergent extraction of myelin proteins generates potent inhibitory fractions [8]. The final product contains MAIs and residual lipids. It is important to avoid freeze/thawing of myelin and to always keep the diluted samples on ice to prevent inactivation. Due to the presence of lipids, it is important to sonicate the stock solution (three times for 5 s at 20 % amplitude with 5 s cooling intervals) before preparing treatments to prevent aggregation. When analyzing dose curves of myelin membranes, be aware that neurons fail to adhere to high concentrations of myelin substrates.

Nogo-A: This MAI has been extensively used to inhibit axonal outgrowth. Several peptides derived from sequences in Nogo-A can exhibit inhibitory activity [3, 15, 16]. We have used the Nogo-66 fragment for neurite outgrowth assays and growth cone collapse assays. Nogo-66 can be isolated from bacteria and heterologous cells using affinity purification for glutathione-S-transferase (GST) or alkaline-phosphatase (AP) fusion proteins and clustered to optimize its activity [17].

Myelin-associated glycoprotein (MAG): MAG has been commonly used to assess neurite outgrowth inhibition and can be coated onto the culture dish or expressed on the cell surface of heterologous cells. Clustering of MAG is often critical for optimal activity [18].

3.2 DRG Dissection

1. Sterilize dissection instruments and dissecting hood with 70 % ethanol. Wait until the instruments are completely free of any residual ethanol before beginning the dissection.
2. Crack the eggshell at the level of the air cell using the bottom of the forceps. Remove the inner shell membrane to reach the embryo inside the egg.
3. Remove the embryo and sacrifice it by decapitating it with spring scissors. Transfer the body to a clean petri dish and place it on its back, ventral side up. Using the forceps, cut the abdominal and thoracic tissues parallel to the spinal column, exposing the viscera (Fig. 1a).

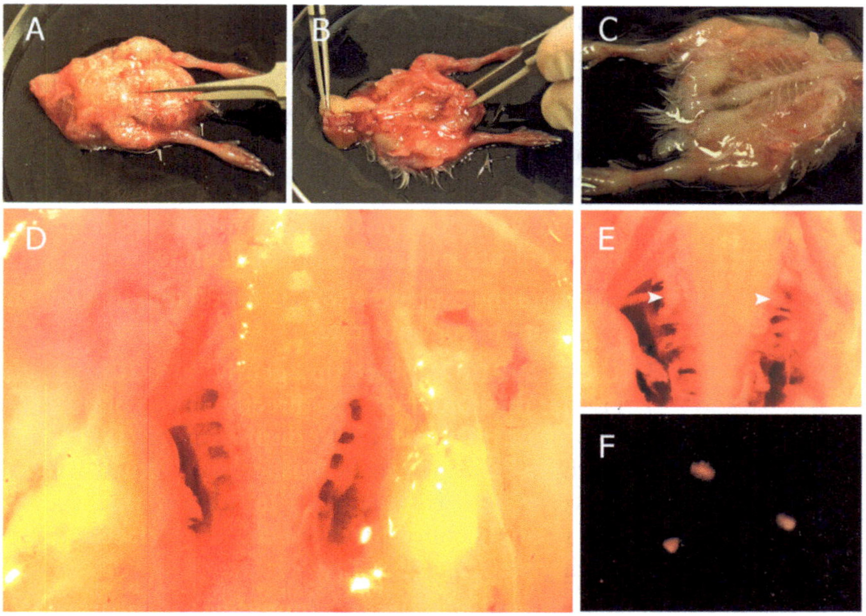

Fig. 1 Dissection of DRGs from an E14 chick embryo. (**a**) Opening of the abdominal and thoracic cavities. (**b**) Removal of the viscera. (**c**) Aspect of the eviscerated embryo. (**d**) Detail of the lumbosacral spinal column of an eviscerated embryo. (**e**) Lumbosacral spinal column after the removal of the meninges. *Arrowheads*: Exposed DRGs. (**f**) DRGs after the complete removal of the surrounding tissue and both its dorsal and ventral roots

4. Under the microscope, remove the visceral organs of the embryo (Fig. 1b). Using the disposable transfer pipets, occasionally rinse the embryo with L15 medium to wash away debris and to prevent the tissue from drying (Fig. 1c, d). Using the forceps, carefully remove the dorsal tissue and meninges to expose the lumbosacral DRGs (Fig. 1e) (*see* **Note 4**).
5. Using a pair of forceps, restrain the spinal nerve while using the other pair to extract the DRG. DRGs can be easily removed by pinching the dorsal root between the ganglia and the spinal cord. Transfer the DRG to a clean petri dish containing L15 medium (Fig. 1f).
6. Remove any remnants of the spinal nerve and surrounding tissue to isolate the DRG (Fig. 1f).

3.3 Growth Cone Collapse Assay

3.3.1 DRG Explant Culture

All steps described here are designed for cultures using 4-well culture dishes. The same procedure can be followed for other formats. Chick DRGs can be isolated and cultured from E6 to E16 under the same conditions. DRG explants from younger embryos extend axons faster and contain less non-neuronal cells than older embryos.

1. Cut DRG explants into two to four pieces with a blade or microdissection scissors. Some DRGs are very small, but it is important to cut them into pieces to expose the DRG neurons to the substrate.

2. Remove PBS from the wells and add 100 μL growth medium, ensuring that the medium fully covers the bottom of the well but is not in excess. For optimal adherence of the neurons to the substrate, the medium should bathe the explant to prevent it from drying, but should not support flotation of the explant.
3. Lay down three to four DRG fragments in a line in the center of each well. Confirm that the explants are in contact with the tissue culture substrate and not floating.
4. Incubate the explants at 37 °C for 30 min to allow the explants to adhere to the substrate.
5. Using a P1000 pipet tip, gently add 200 μL equilibrated culture medium dropwise directly on top of the explants. This will cause minimal DRG displacement. If an explant floats to the surface, it can be pushed back down with a pipet tip; however, this is usually an indication that the explant did not strongly contact the cell culture surface during the initial plating step.
6. Place chamber slides in a humidified incubator at 37 °C with 5 % CO_2 for 12–18 h. The incubation time varies with the developmental stage of the DRGs; however, during the initial stages of outgrowth (<18 h), the growth cones are maximally expanded and have clearly distinguishable actin-rich lamellipodia and filopodia. Older cultures tend to have extended axons with bullet-shaped growth cones.

3.3.2 Growth Cone Collapse (See Fig. 2)

Growth cones in culture can collapse in response to changes in temperature and pH. We have highlighted critical steps to minimize basal growth cone collapse.

1. Prepare fixative containing 20 % sucrose immediately before treating DRG explants.
2. Dilute treatments (e.g., Nogo, MAG) in growth medium to a concentration of 3× at a final volume of 100 μL and equilibrate the treatments in the incubator to 37 °C and 5 % CO_2.
3. In the tissue culture incubator, carefully remove 100 μL medium from each well. Place the pipet tip at the surface of the meniscus against the side of the well to remove the medium. Removing 1/3 of the volume will not disturb the explants and growth cones.
4. Add 100 μL 3× treatment dropwise to corresponding wells. Gently rock the chamber slide to mix, and incubate for the appropriate time. For axonal repellents such as Semaphorin 3A, robust collapse usually occurs within 20 min, while MAIs generally lead to significant collapse in 60 min [6, 19]. (*See* **Note 6** for application of soluble myelin).

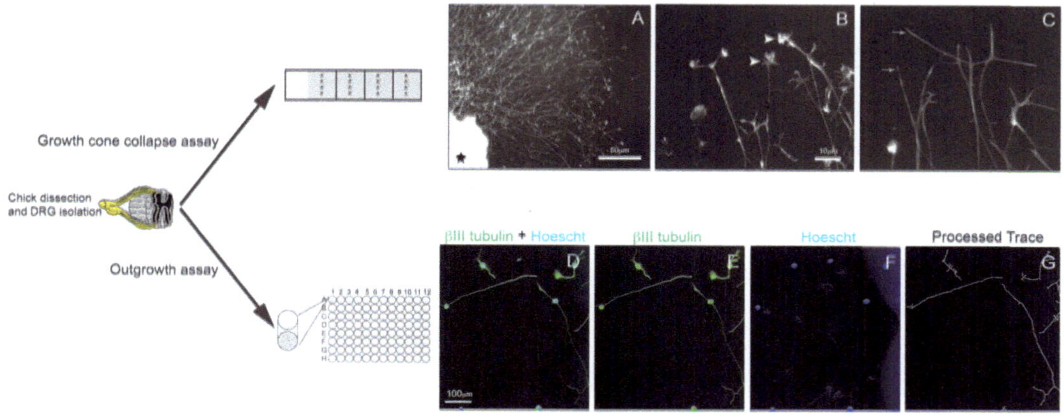

Fig. 2 Outline of neurite outgrowth and growth cone collapse procedures. DRGs are dissected and dissociated. The neurons are then seeded as explants in 4-well chamber slides for the growth cone collapse assay (*top*) or in 96-well plates for the neurite outgrowth assay (*bottom*). For growth cone collapse, explants are stained with rhodamine phalloidin (**a**) and imaged at high power (**b**, **c**) to image spread (*arrowheads*, **b**) or collapsed (*arrows*, **c**) growth cones. For neurite outgrowth assays, cultures are stained with anti-βIII tubulin antibody and Hoescht nuclear stain (**d–f**). Images can be processed using programs such as the NeuriteTracer ImageJ plugin or MetaXpress software to define the length of processes derived from βIII tubulin/Hoescht—double positive cells (**g**)

5. Remove the cells from the incubator and immediately add 300 μL fixative, placing the pipet tip at the bottom of the well against the side. The sucrose makes the fixative dense and it will layer underneath the tissue culture medium when added gently to the bottom of the well.
6. After 30 min of fixation at room temperature, gently wash each well three times 10 min with room temperature PBS.

3.3.3 Analysis and Quantification of Growth Cone Collapse

Negative cues tend to collapse growth cones, whereas positive cues expand them (Fig. 2a–c). An efficient marker of growth cone morphology is fluorescently conjugated phalloidin, which binds to filamentous actin [20]. We use rhodamine-phalloidin to stain growth cones, which can be easily identified at the tips of neuritic processes projecting in a radial fashion away from the DRG explant.

1. *Staining*
   (a) Permeabilize fixed cultures with 0.2 % Triton X-100 in PBS for 5 min. Wash once with PBS.
   (b) Incubate with rhodamine-phalloidin for 1 h at room temperature. Remove phalloidin and wash three times 10 min with PBS.
   (c) Peel off the wells and adhesive on the chamber slide and mount a cover slip on the slide with a mounting medium that includes an anti-fade reagent such as mowiol with *p*-phenylenediamine.

2. *Quantification*: Growth cone morphologies can be grouped into several categories. Traditionally, growth cones with any visible filopodia and lamellipodia are considered expanded, while bullet-shaped growth cones are counted as collapsed [6] (Fig. 2). Image the growth cones with an epifluorescence microscope, using a 20× or 40× objective. Growth cones can be rapidly scored as spread or collapsed with a counter at the microscope. Quantify duplicates of each condition to account for variability of explant growth between wells. It is also important to quantify more than one explant per well. Growth cone collapse index is commonly represented by the percent of collapsed growth cones in the total population.

3.4 Outgrowth Assay (Fig. 2)

Dissociated cultures are preferable for quantification of neurite outgrowth to fully assess the extent of growth from the cell body. To reliably analyze outgrowth per neuron, it is essential to plate the DRG neurons at a low density to minimize intercellular contacts.

3.4.1 Dissociation of the DRGs (See Note 2 for Rodent Modifications)

1. Transfer whole DRG explants to a 15 mL conical tube and centrifuge for 3 min at 1,000 × *g*.
2. Remove the medium using a pasteur pipet and add 2 mL 0.25 % Trypsin-EDTA. Incubate the DRGs in trypsin–EDTA for 30 min in a 37 °C, 5 % CO_2 incubator.
3. Centrifuge the digested DRGs for 3 min at 1,000 × *g*, remove the trypsin, and resuspend the DRGs with 5 mL culture medium containing 10 % FBS to neutralize residual trypsin.
4. Centrifuge the DRGs for 3 min at 1,000 × *g* and carefully remove the supernatant.
5. Add 1 mL growth medium and triturate the DRGs by gently pipetting up and down with a P1000 pipet tip until the solution becomes homogeneously turbid (about 10 times). Excessive trituration will reduce neurite outgrowth and viability.
6. Dilute the cell suspension in growth medium to a final volume of 5 mL, allow debris to settle to the bottom of the tube, and remove the supernatant containing dissociated DRG neurons.
7. Count the cells with a hemocytometer and plate at a density appropriate to the experimental application. Cell densities around 15.5×10^3 cells/cm^2 typically generate healthy cultures in which the confluency does not impair assessment of neurite outgrowth.
8. Place the plates in a humidified incubator at 37 °C with 5 % CO_2 for 12–18 h.

Optional Step: Enrichment of Neuronal Content

In order to reduce glial content in these cultures, the following procedures can be followed.

1. Culture the cell suspension in a 100 mm tissue culture dish for 1 h to allow the non-neuronal cells to adhere to the dish.
2. Collect the medium containing unattached neurons and transfer to a 15 mL conical tube. Centrifuge at 1,000 × *g* for 5 min.
3. Aspirate the supernatant and resuspend the cell pellet with 1 mL growth medium.
4. Count the cells with a hemocytometer and plate the cells at a density appropriate to the experimental application.

3.4.2 Analysis of Neurite Outgrowth

1. *Immunocytochemistry*
   (a) Fix the cells with ice-cold 4 % PFA in PBS containing 20 % sucrose for 30 min. Wash three times for 5 min with PBS to remove any residual PFA.
   (b) Permeabilize the cells for 5 min with 0.2 % Triton X-100 in PBS and wash three times with PBS.
   (c) To reduce nonspecific binding, block with 5 % BSA in PBS at room temperature for 1 h.
   (d) Remove the BSA and incubate with anti-BIII-TUJ1 tubulin (1:2,000) in 1 % BSA overnight at 4 °C to label the neuronal processes.
   (e) Wash three times for 5 min with PBS and incubate with fluorophore-conjugated secondary antibody (1:2,000), and a nuclear stain (Hoechst/DAPI) (1:2,000) in 1 % BSA for 1 h at room temperature.
   (f) Wash three times for 5 min with PBS and image the stained cultures with a fluorescence microscope.
2. *Quantification*: The ImageXpress System (Molecular Devices) is a microscope that automates the acquisition and analysis of fluorescent images. Total neurite length, mean neurite length, and number of branches per cell can be assessed using the Neurite Outgrowth module of the MetaXpress software for the ImageXpress System, allowing for high-throughput analysis of neuronal cultures. The user can establish thresholds for shape, size, and intensity. Alternatively, our group has developed a freely available ImageJ plugin, called NeuriteTracer [21], for automated tracing and measurement of neurites derived from βIII tubulin/Hoescht—double positive cells (Fig. 2d–g). For consistency, during the acquisition of images, regions of interest should be restricted to the center of the well (typically 12–18 sites) because the periphery can contain a higher density of interneuronal contacts, interfering with the quantification of individual neurites. As with the growth cone collapse assay, it is suggested to have duplicate or triplicate wells for each condition.

3.5 Viral-Mediated Delivery in DRG Neurons

DRG neurons are a good model to study outgrowth in response to experimental manipulation of gene expression because they are readily infected by a number of viral vectors. Although dissociated DRG neurons can be electroporated or transfected, these approaches are not efficient for explants. Viral transduction is an efficient way to mediate protein overexpression or knockdown in dissociated cultures and explants. We find that lentivirus and herpes simplex virus type 1 (HSV-1) infect DRG neurons with close to 100 % efficiency. We find that infection with replication-deficient recombinant HSV yields efficient and robust overexpression, with significant expression of fluorescently tagged proteins after only 3–8 h [22]. Neurite outgrowth and growth cone collapse assays can be performed at least 24 h after infection [17, 23].

3.6 Conclusion

Neurite outgrowth and growth cone collapse assays using cultured DRG neurons provide a powerful system to study molecular processes involved in neurite growth in development and regeneration contexts. We have offered a number of approaches to study the molecular basis of neurite outgrowth inhibition and growth cone collapse in response to repellent molecules in the adult CNS, with an emphasis on myelin, individual MAIs, and CSPGs. The protocols described herein can be used to establish neurite outgrowth inhibition and growth cone collapse in culture, with the goal of screening for the ability of agents to relieve this inhibition, thereby allowing for the identification of molecular regulators of axon regeneration.

4 Notes

1. Rodent DRGs can also be used to assess growth cone morphology and neurite outgrowth; however growth cones are smaller. A number of visual references for rodent DRG dissection can be found in the literature [24, 25]. Chick DRGs become responsive to inhibitors in the adult CNS at approximately E12, whereas rodent DRG neurons become responsive at P4. E6–E8 chick DRGs and late embryonic or early postnatal rodent DRGs can be cultured to assess responses to developmental guidance cues [7].
2. Rodent DRG dissociation should be performed with 0.25 % Trypsin-ETA and 0.1 % Collagenase Type 1A-S. For dissociating rodent DRGs, trituration can be performed with a syringe and three different gauge needles (21-23-25G).
3. Optimal growth medium for rodent cultures is Neurobasal medium supplemented with 2 % B27, 50 ng/mL NGF, penicillin/streptomycin, and 0.5 nM glutamine.
4. Careful dissection of the dorsal tissue may preserve the sympathetic trunk, which is a thin, flat structure that lies directly

ventral to (on top of) the chain of DRGs. It is important to carefully remove this structure to expose the underlying DRGs.

5. Drying down protein can lead to nonspecific neurite outgrowth inhibition. We find that drying down myelin membranes is the best method to assess the sum impact of multiple myelin inhibitors. An excellent control for dried myelin membranes is heat-inactivated myelin [26]. It is preferable to apply purified recombinant proteins through coating rather than drying; however, dried bovine serum albumin (BSA) or tags such as glutathione-S-transferase (GST), where appropriate, can serve as a control for drying down recombinant proteins if necessary [27].
6. As the DRG growth medium does not contain serum, some aggregation of myelin membranes may lead to variable effects across the well. This can be avoided by adding a small amount of serum to the treatment medium when performing growth cone collapse; 0.1–0.2 % FBS limits aggregation without causing growth collapse on its own.

Acknowledgments

A.E.F. is a Canada Research Chair and R.A. is supported by a Lloyd Carr-Harris Studentship. This work is supported by the McGill Program in Neuroengineering and by grants from the Canadian Institutes of Health Research and the MS Society of Canada.

References

1. Kaech S, Banker G (2006) Culturing hippocampal neurons. Nat Protoc 1(5):2406–2415. doi:10.1038/nprot.2006.356
2. Yiu G, He Z (2006) Glial inhibition of CNS axon regeneration. Nat Rev Neurosci 7(8):617–627. doi:10.1038/nrn1956
3. Chen MS, Huber AB, van der Haar ME, Frank M, Schnell L, Spillmann AA, Christ F, Schwab ME (2000) Nogo-A is a myelin-associated neurite outgrowth inhibitor and an antigen for monoclonal antibody IN-1. Nature 403(6768):434–439. doi:10.1038/35000219
4. Koprivica V, Cho KS, Park JB, Yiu G, Atwal J, Gore B, Kim JA, Lin E, Tessier-Lavigne M, Chen DF, He Z (2005) EGFR activation mediates inhibition of axon regeneration by myelin and chondroitin sulfate proteoglycans. Science 310(5745):106–110. doi:10.1126/science.1115462
5. Prinjha R, Moore SE, Vinson M, Blake S, Morrow R, Christie G, Michalovich D, Simmons DL, Walsh FS (2000) Inhibitor of neurite outgrowth in humans. Nature 403(6768):383–384. doi:10.1038/35000287
6. Raper JA, Kapfhammer JP (1990) The enrichment of a neuronal growth cone collapsing activity from embryonic chick brain. Neuron 4(1):21–29
7. Cai D, Qiu J, Cao Z, McAtee M, Bregman BS, Filbin MT (2001) Neuronal cyclic AMP controls the developmental loss in ability of axons to regenerate. J Neurosci 21(13):4731–4739
8. Norton WT, Poduslo SE (1973) Myelination in rat brain: method of myelin isolation. J Neurochem 21(4):749–757

9. Nash M, Pribiag H, Fournier AE, Jacobson C (2009) Central nervous system regeneration inhibitors and their intracellular substrates. Mol Neurobiol 40(3):224–235. doi:10.1007/s12035-009-8083-y
10. Busch SA, Silver J (2007) The role of extracellular matrix in CNS regeneration. Curr Opin Neurobiol 17(1):120–127. doi:10.1016/j.conb.2006.09.004
11. Kwok JC, Afshari F, Garcia-Alias G, Fawcett JW (2008) Proteoglycans in the central nervous system: plasticity, regeneration and their stimulation with chondroitinase ABC. Restor Neurol Neurosci 26(2–3):131–145
12. Sharma K, Selzer ME, Li S (2012) Scar-mediated inhibition and CSPG receptors in the CNS. Exp Neurol 237(2):370–378. doi:10.1016/j.expneurol.2012.07.009
13. Ahmed Z, Mazibrada G, Seabright RJ, Dent RG, Berry M, Logan A (2006) TACE-induced cleavage of NgR and p75NTR in dorsal root ganglion cultures disinhibits outgrowth and promotes branching of neurites in the presence of inhibitory CNS myelin. FASEB J 20(11):1939–1941. doi:10.1096/fj.05-5339fje
14. Alabed YZ, Pool M, Ong Tone S, Fournier AE (2007) Identification of CRMP4 as a convergent regulator of axon outgrowth inhibition. J Neurosci 27(7):1702–1711. doi:10.1523/JNEUROSCI.5055-06.2007
15. Huebner EA, Kim BG, Duffy PJ, Brown RH, Strittmatter SM (2011) A multi-domain fragment of Nogo-A protein is a potent inhibitor of cortical axon regeneration via Nogo receptor 1. J Biol Chem 286(20):18026–18036. doi:10.1074/jbc.M110.208108
16. GrandPre T, Nakamura F, Vartanian T, Strittmatter SM (2000) Identification of the Nogo inhibitor of axon regeneration as a Reticulon protein. Nature 403(6768):439–444. doi:10.1038/35000226
17. Hsieh SH, Ferraro GB, Fournier AE (2006) Myelin-associated inhibitors regulate cofilin phosphorylation and neuronal inhibition through LIM kinase and Slingshot phosphatase. J Neurosci 26(3):1006–1015. doi:10.1523/JNEUROSCI.2806-05.2006
18. Domeniconi M, Zampieri N, Spencer T, Hilaire M, Mellado W, Chao MV, Filbin MT (2005) MAG induces regulated intramembrane proteolysis of the p75 neurotrophin receptor to inhibit neurite outgrowth. Neuron 46(6):849–855. doi:10.1016/j.neuron.2005.05.029
19. Ferraro GB, Morrison CJ, Overall CM, Strittmatter SM, Fournier AE (2011) Membrane-type matrix metalloproteinase-3 regulates neuronal responsiveness to myelin through Nogo-66 receptor 1 cleavage. J Biol Chem 286(36):31418–31424. doi:10.1074/jbc.M111.249169
20. Wulf E, Deboben A, Bautz FA, Faulstich H, Wieland T (1979) Fluorescent phallotoxin, a tool for the visualization of cellular actin. Proc Natl Acad Sci U S A 76(9):4498–4502
21. Pool M, Thiemann J, Bar-Or A, Fournier AE (2008) NeuriteTracer: a novel ImageJ plugin for automated quantification of neurite outgrowth. J Neurosci Methods 168(1):134–139. doi:10.1016/j.jneumeth.2007.08.029
22. Neve RL, Lim F (2013) Generation of high-titer defective HSV-1 vectors. Curr Protoc Neurosci Chapter 4:Unit 4 13. doi:10.1002/0471142301.ns0413s62. Editorial board, Jacqueline N Crawley [et al.]
23. Alabed YZ, Pool M, Ong Tone S, Sutherland C, Fournier AE (2010) GSK3 beta regulates myelin-dependent axon outgrowth inhibition through CRMP4. J Neurosci 30(16):5635–5643. doi:10.1523/JNEUROSCI.6154-09.2010
24. Saijilafu, Zhou FQ (2012) Genetic study of axon regeneration with cultured adult dorsal root ganglion neurons. J Vis Exp (66). doi:10.3791/4141
25. Albuquerque C, Joseph DJ, Choudhury P, MacDermott AB (2009) Dissection, plating, and maintenance of dorsal root ganglion neurons for monoculture and for coculture with dorsal horn neurons. Cold Spring Harb Protoc 2009(8):pdb.prot5275. doi:10.1101/pdb.prot5275
26. Pool M, Niino M, Rambaldi I, Robson K, Bar-Or A, Fournier AE (2009) Myelin regulates immune cell adhesion and motility. Exp Neurol 217(2):371–377. doi:10.1016/j.expneurol.2009.03.014
27. Fournier AE, Gould GC, Liu BP, Strittmatter SM (2002) Truncated soluble Nogo receptor binds Nogo-66 and blocks inhibition of axon growth by myelin. J Neurosci 22(20):8876–8883

Chapter 5

Quantitative Assessment of Neurite Outgrowth in Mouse Retinal Explants

Tom Buyens, Djoere Gaublomme, Inge Van Hove, Lies De Groef, and Lieve Moons

Abstract

Despite intensive research efforts over the past years, regeneration of injured axons in the central nervous system (CNS) remains elusive. The discovery of novel neuro-stimulatory agents that promote regeneration is hampered by a gap between high content analysis platforms using neuronal cells and time-consuming preclinical animal models. In this regard, tissue explant cultures, which are easily manageable and more closely resemble the in vivo situation, form an ideal model system to study the effect of compounds on the neuroglial network. Retinal explants have proven to be a useful tool to investigate the effect of molecules on neuronal survival and regeneration. In this chapter, we report a detailed description of how to isolate and culture retinal explants and how to immunolabel the outgrowing neurites. Furthermore, we describe different analysis tools, both manual and automated, to quantify neurite outgrowth from retinal explants.

Key words Tissue explant, Retina, Neurite outgrowth, Axonal regeneration, Automated morphometric analysis

1 Introduction

Due to the increasing life expectancy, neurodegenerative diseases, such as Alzheimer's and Parkinson's diseases, age-related macular degeneration and glaucoma, are more prevalent than ever before. Therefore, the development of novel therapies for such neurodegenerative diseases and for central nervous system (CNS) injuries in general has become a top priority, as CNS damage drastically diminishes the quality of life and current treatments are inadequate. For many years, the prevailing dogma held that the adult mammalian CNS is unable to regenerate injured or diseased axons. This lack of spontaneous axonal regeneration of adult neurons has been attributed to a loss of neurotrophic support from target neurons, leading to an impaired neuronal survival [1], an inhibitory environment, which prevents axonal regrowth [2], and a decreased growth potential as compared to embryonic neurons [3].

Andrew J. Murray (ed.), *Axon Growth and Regeneration: Methods and Protocols*, Methods in Molecular Biology, vol. 1162, DOI 10.1007/978-1-4939-0777-9_5, © Springer Science+Business Media New York 2014

The optical circuit forms an ideal model system to study neuronal survival and axonal regeneration in the diseased or injured CNS, because of its well-characterized morphology and functionality. Furthermore, all axons in the optic nerve (ON) arise from a single neuronal cell population, the retinal ganglion cells (RGCs), which can be experimentally manipulated with minimal surgery. Recent in vivo studies have shown that adult RGCs can be stimulated to overcome axonal injury, to extend their axons past the lesion site and to reinnervate their original targets [4, 5]. This discovery has shattered the dogma and formed new hope for patients affected by neurodegenerative disorders or CNS injuries, such as spinal cord injury, and awoken an increased interest of the pharmaceutical industry in the discovery of new therapeutic strategies that stimulate neuroprotection and regeneration [4].

Efforts to identify novel compounds for regenerative treatments are limited by a gap between high-throughput screening models and validation in preclinical animal models. High-content and high-throughput screening methods are mostly based on dissociated cells to assess the effect of a compound, thereby enabling the screening of huge compound libraries and the rapid identification of novel lead compounds. Unfortunately, dissociated cells can show an artificial behavior and they are often only used to elucidate a direct effect on the cells in culture. Preclinical animal models, on the other hand, maintain the intercellular relationship, thereby closely resembling the actual pathology. However they are less efficient and time consuming, often because it can take several weeks before the actual disease manifests. Together with the high degree of variability between animals, this results in a costly and labor intensive experimental setup. Organotypical culture models, such as retinal explants, form an ideal bridge to close this gap, as they provide the necessary compromise between the two model systems. The major advantage of explant cultures is that cells are kept within their normal environment, thereby preserving cell-to-cell interactions, while maintaining a higher level of experimental control as in animal models [6]. Organotypic cultures thereby provide an ideal platform for the identification and validation of novel neuroprotective or pro-regenerative substances. As retinal explants have previously been used to study various neural processes such as retinal differentiation [7, 8], neurodegeneration [9, 10], and neurite outgrowth [11–13], they provide an ideal ex vivo system to screen promising molecules.

Neurite outgrowth can be visualized by fluorescent immunohistochemistry (IHC) and optical microscopy, such as phase contrast and differential interference contrast (DIC) microscopy. These optical microscopic techniques are often used in combination with manual analyses. As no prior treatment is required, explants can be studied while they are still in culture, thereby allowing time-lapse experiments. A disadvantage is that no objective distinction can be made between glial and neuronal processes, which may result in a false representation of neurite outgrowth. This distinction can be made using antibodies targeting neuronal markers, such as β-tubulin or neurofilament. Neurofilament is frequently used as a neurite marker in retinal cell and explant cultures of adult animals, resulting in a clear and uniform staining. In neonatal cultures, however, a neurofilament immunostaining results in a patchy expression pattern along RGC neurites, making it unsuitable for neurite quantification [14]. β-Tubulin, on the other hand, is commonly used as a marker for RGCs and their axons, providing a uniform staining for neurite outgrowth both in adult and in neonatal cultures. The explant body can easily be visualized using the nuclear marker DAPI. The main disadvantage of using immunological stainings is that time-lapse experiments are no longer possible afterwards. Of note, transgenic animals, that express a fluorescent protein in RGC axons, such as the thy1-YFP mice [15], might allow for time-lapse experiments. However, axonal injury leads to a large shift in the expression pattern of RGCs, e.g., a decreased expression of thy1, thereby resulting in a large decrease in GFP expression in newly formed neurites and thus an underestimation of neurite outgrowth (personal observations and refs. [16, 17]).

Neurite outgrowth assessment in explant cultures is typically done using manual or semiautomated analyses. Manual tracing of "neurite length" (Fig. 1a) and quantification of the "neurite outgrowth area" (NOA) (Fig. 1b) are two of the most common methods to quantify neurite outgrowth [12, 18]. However, these analyses are time consuming, labor intensive, and prone to user-biased errors. Therefore, we recently developed a custom made analysis script to automatically quantify the "immunodetected neurite area" (INA) (Fig. 1c) [19]. Dual-colored pictures are split into two channels, one containing the explant body as visualized by DAPI staining and the second channel representing the network of immunolabeled neurites. Morphological features in both channels are detected by local thresholding and subsequent filtering steps, resulting in a clear detection of, respectively, the explant and its outgrowing neurites. The INA is then quantified as the area that is occupied by the detected neurites, normalized to the perimeter of the explant. Furthermore, the INA can also be categorized into four segments by drawing three concentric rings, generated by successive 100 μm increments from the explant body edge and categorizing neurite outgrowth in three inner ring segments and an outer segment with the remaining detected neurites (Fig. 1d). This categorization can be used to study compound-induced effects on axonal extension, by comparing the neurite area in the outer segment, which only contains long axons, relative to that of the inner segment, representing all axons. Furthermore, we recently included an automated NOA analysis based on the

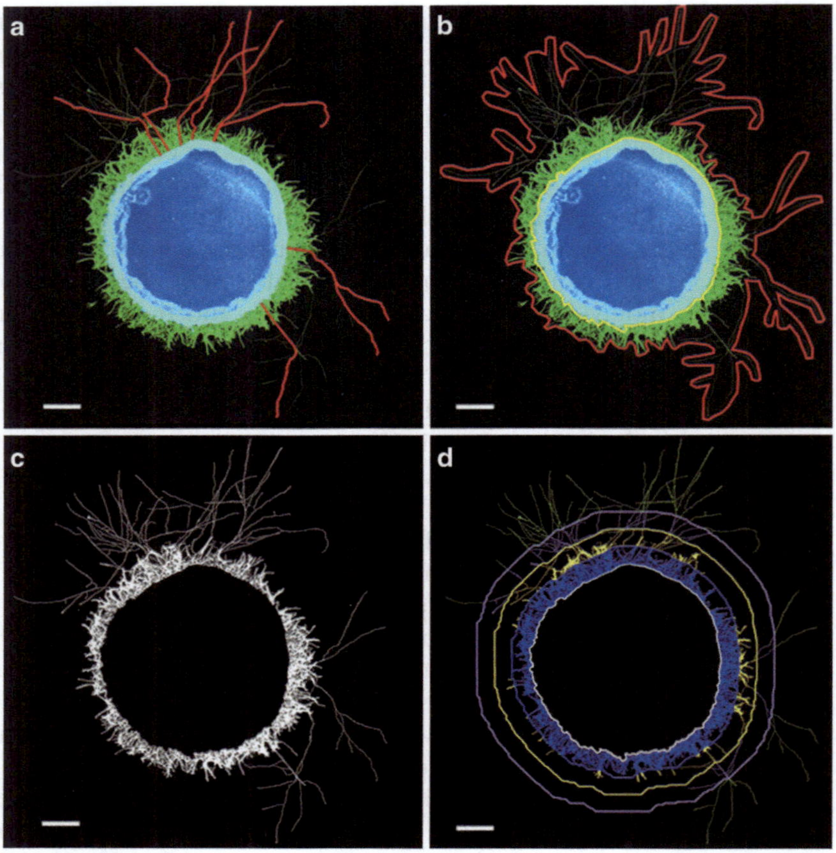

Fig. 1 Overview of quantification methods determining neurite outgrowth in retinal explants. (**a**) In the "maximum neurite length" analysis method, the 15 longest neurites are manually traced as shown in *red*. (**b**) In the "Neurite outgrowth area" or NOA analysis method, the total neurite area is outlined, as shown in *red*. The area of the explant body, outlined in *white*, is then subtracted. (**c**, **d**) The automated analysis method automatically detects the "immunopositive neurite area" or INA (*white*) (**c**) and segments it into four concentric rings, shown in different colors (*blue*, *purple*, *yellow*, and *green*) (**d**). Figure modified from ref. [19]

detected neurites from the INA assay. In this analysis, the detected neurites are dilated and any remaining gaps in the dilated area are filled, thereby marking the neurite outgrowth area, similar as in the manual NOA analysis. The use of this automated script offers many advantages, such as increased objectivity and reproducibility. Most importantly, the speed of analysis is significantly higher compared to the manual methods, which take several hours hands-on time per experiment.

Taken together, retinal explants form an interesting model to efficiently evaluate neurite outgrowth in an in vivo-like situation. This chapter precisely describes retinal explant dissection, culture, immunostaining, and analysis methods. Of note, all analysis tools are described based on Axiovision software, but other analysis programs, such as ImageJ, can also be used after mild adaptations.

2 Materials

2.1 Tissue Dissection and Retinal Explant Culture

All dissection tools should be sterilized with 70 % ethanol or autoclaved and all solutions should be prepared with ultrapure water and autoclaved.

1. Dissection microscope with light source, e.g., Leica S6E.
2. Dissection instruments: Blunt standard scissors, curved Dumont #7 forceps, 2 Dumont #5 forceps, 23G needle, Vannas spring scissor, Blunt straight forceps, and Versi handle with 750 μM FUE standard punch needle (Mediquip Surgical).
3. P200 micropipette.
4. Ice-cold Neurobasal-A medium (Invitrogen).
5. Phosphate-buffered saline (PBS) consisting of 8 mM $Na_2HPO_4.2H_2O$, 2 mM $KHPO_4$, 150 mM NaCl, and 3 mM KCl, pH 7.4.
6. Poly-L-lysine hydrobromide (M_W 70,000–150,000, Sigma) diluted in PBS to a concentration of 0.25 mg/mL.
7. Laminin from Engelbreth-Holm-Swarm murine sarcoma basement membrane (Sigma) diluted in PBS to a concentration of 2 μg/mL.
8. Nunc round 4-well culture plates (Thermo Scientific).
9. Explant culture medium consisting of Neurobasal-A medium supplemented with 1 mM L-glutamine, 0.25 μg/mL fungizone, 100 U/mL penicillin, 100 μg/mL streptomycin, 2 % B27 supplement (all from Invitrogen), and 0.4 % methylcellulose (Sigma). Culture medium is stored at 4 °C in the dark (*see* **Note 1**).
10. Mice: 6–8 littermate pups (at 0–3 days after birth (*see* **Note 2**)).

2.2 Immunohistochemistry

1. Fixation solution consisting of 4 % paraformaldehyde in PBS (see above).
2. Washing buffer containing 50 mM Tris, 150 mM NaCl, and 0.1 % Triton X-100, pH 7.6.
3. Incubation buffer containing 50 mM Tris, 150 mM NaCl, and 0.5 % blocking reagent (Perkin Elmer), pH 7.6.
4. Blocking solution containing 20 % pre-immune serum of goat (Invitrogen) in incubation buffer.
5. Primary antibody solution containing monoclonal mouse anti-β-tubulin antibody (Sigma, clone SDL.3D10) diluted 1/500 in incubation buffer.
6. Secondary antibody solution containing goat anti-mouse IgG labeled with Alexa Fluor 488 (Invitrogen) diluted 1/500 in incubation buffer (*see* **Note 2**).

7. DAPI solution consisting of 1 μg/mL DAPI (Applichem) in PBS.
8. Storage buffer consisting of 0.01 % sodium azide in PBS.

2.3 Image Acquisition and Analysis

1. Image acquisition is performed using an inverted confocal microscope, e.g., FV1000 IX-82 microscope (Olympus).
2. Image analysis is performed using Axiovision analysis software (Zeiss).

3 Methods

The following procedures are all performed under a horizontal laminar flow to prevent contamination.

3.1 Coating of Cell Culture Plates

1. The day before dissection, culture plates are coated with poly-L-lysine. Fill each well with 300 μL of 0.25 mg/mL poly-L-lysine solution and place the plates in the incubator at 37 °C for overnight incubation (*see* **Note 3**).
2. On the day of dissection, aspirate the poly-L-lysine solution and rinse the plates three times with sterile PBS.
3. Fill the wells with 300 μL laminin solution and incubate for 2 h in the incubator at 37 °C.
4. Aspirate the laminin solution and rinse twice with PBS.
5. Add 250 μL of explant culture medium to each well and store in the incubator at 37 °C till use.

3.2 Explant Dissection

1. Euthanize mouse pups by decapitation (*see* **Note 4**).
2. Disinfect the head with 70 % ethanol spray.
3. Peel off the skin covering the eye using Dumont #5 forceps, carefully dissect the eyeball using the Dumont #7 curved forceps, and place it in a drop of ice-cold Neurobasal medium in a plastic petri dish (*see* **Note 5**). Store one eye on ice, while dissecting the other under a dissection microscope.
4. Punch a hole in the limbus with the 23 G needle, while holding the eye by the ocular muscles (Fig. 2a, arrow). Cut along the limbus using the spring scissors to remove the anterior chamber (Fig. 2b). Remove the lens, inner limiting membrane, and blood vessels from the retinal cup using Dumont #5 forceps. Finally remove the retinal pigment epithelium/sclera and pinch off the optic nerve. Make a perpendicular cut from de edge of the retina to the optic nerve head (Fig. 2c).
5. Fix the retinal cup by placing the blunt forceps at the optic nerve head. Place the punch needle at the periphery of the

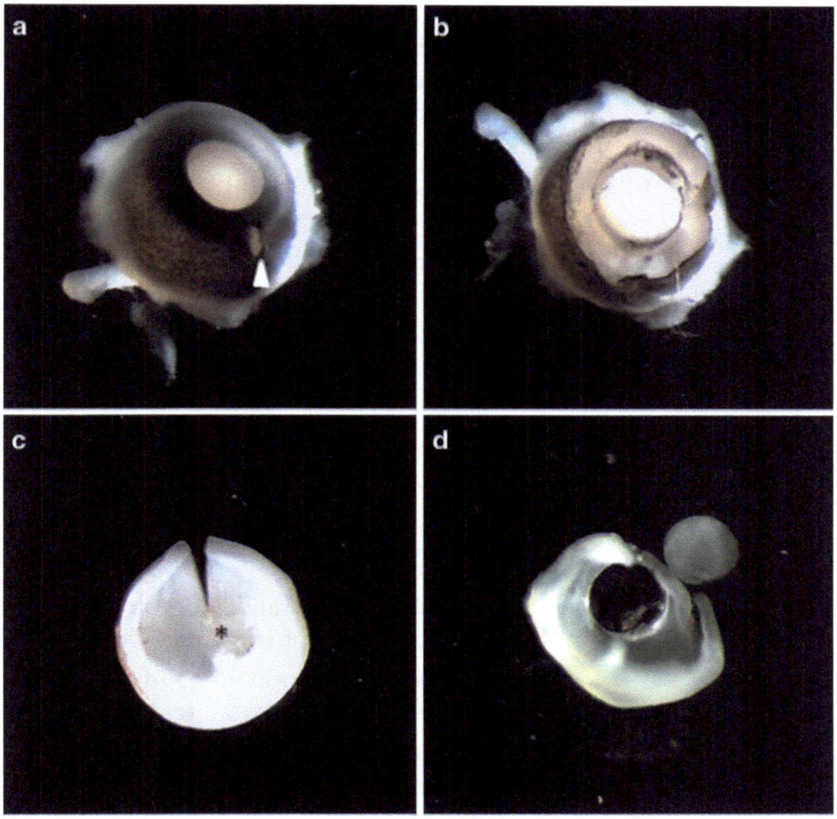

Fig. 2 Overview of the dissection of retinal explants. A small hole (*arrow*) is punched at the limbus of an isolated eye using a 23 G needle (**a**). The anterior chamber is cut off at the limbus (**b**), next the lens and pigmented sclera are removed and an incision is made from the edge of the retina to the optic nerve head as indicated by (*asterisk*) (**c**). Explants are excised one by one at a similar peripheral position, while fixing the retina by placing a blunt forceps at the optic nerve head (**d**)

retina and gently press down to excise the explant (Fig. 2d) (*see* **Notes 6** and 7). Most of the time, the punched explants will immediately attach to the petri dish, however if the explant sticks in the punch needle, gently dislodge it using fine forceps.

6. Turn the eye cup somewhat to the right and continue the dissection until all explants are harvested. Do the same with the other eye. In total 12–16 explants can be harvested from one mouse depending on its age (*see* **Note** 7).
7. Collect the punches using a P200 micropipette with a cut off tip and transfer to a coated 4-well plate with explant medium. All explants are divided over the 4 wells of the plate, thereby seeding 3–4 punches per well, depending on the age of the mouse.

8. Gently orientate the punches using blunt forceps. Explants should be positioned with the RGC layer facing the coated surface (*see* **Note 8**). Do not scratch the coated well bottom as this will interfere neurite outgrowth. Apply gentle pressure to allow good adhesion to the coated surface.
9. Gently add an additional 250 μL of explant medium to each well.
10. Place the culture plate in the incubator at 5 % CO_2 at 37 °C.
11. During the 72 h incubation period, 250 μL of the medium is changed every 24 h (*see* **Note 9**).

3.3 Immunohistochemistry and Microscopy

1. Collect the explants after 3 days of culture.
2. Gently remove the culture medium.
3. Carefully rinse once with cold PBS.
4. Fix the explants with 4 % PFA for 1 h.
5. Rinse the explants three times for 5 min with washing buffer.
6. Aspirate the washing buffer and add 300 μL of blocking solution. Incubate for 45 min at room temperature (*see* **Note 10**).
7. Remove the blocking solution and add the primary antibody solution. Incubate for 2 h at room temperature.
8. Aspirate the primary antibody solution and rinse the plates three times with washing buffer.
9. Add 300 μL of secondary antibody solution in each well and incubate for 2 h at room temperature. Protect from light by covering the plates with tin foil.
10. Rinse the plates three times for 5 min with washing buffer, while protecting them from light.
11. Remove the washing buffer and incubate for 30 min with DAPI solution at room temperature. Protect from light by covering the plates with tin foil.
12. Remove the DAPI solution and add 500 μL of storage buffer. Wrap the plates in tin foil to protect from light and store at 4 °C until microscopic analysis.
13. The 4-well plates with the retinal explants are allowed to warm to room temperature before microscopic analysis. Explants are visualized under an inverted confocal fluorescence microscope with a motorized stage using a 20× objective (*see* **Note 11**). DAPI, used to visualize the explant body, is excited by a 405 nm laser, and the Alexa 488 fluorophore, used to label the outgrowing neurites, by a 488 nm laser. The explant and its neurites are photographed as a two colored picture using a mosaic acquisition function. All pictures are saved as tiff files.

3.4 Computer-Assisted Quantification of Neurite Outgrowth

Although all analysis methods are described using the Axiovision software, alternative morphometric analysis programs, such as ImageJ, can be applied as well.

3.4.1 Computer-Assisted Quantification of Maximum Neurite Length

1. The picture file is opened in Axiovision software. Maximum neurite length is determined by manual tracing of the length of the neurites using the "curve" function in the Axiovision software (*see* **Notes 12** and **13**). Neurites are traced starting from the growth cone to the explant body, visualized by the DAPI staining.
2. Start with the neurite that appears to be the longest and work your way down until the 15 longest neurites are traced (Fig. 1a).
3. Use the "create table" function to collect all measurements in a table. Save this table as an Excel sheet.
4. Repeat **steps 1–3** till all explants are analyzed.
5. The length of the 10 longest neurites is averaged per explant. This value can now be used to compare the average neurite length under different conditions.

3.4.2 Computer-Assisted Quantification of Neurite Outgrowth Area (NOA)

1. Open the picture file in Axiovision software. Use the "outline" tool to quantify the area and perimeter of the explant body visualized by the DAPI staining.
2. Use the "outline" tool to quantify the neurite area by outlining the neurites as visualized by the β-tubulin staining (Fig. 1b) (*see* **Notes 12** and **13**).
3. Use the "create table" function to collect all measurements in a table. Save this table as an Excel sheet.
4. Repeat **steps 1–3** till all explants are analyzed.
5. The NOA is obtained by subtracting the explant body area from the neurite area and subsequently dividing that value by the explant body perimeter. This value can now be used to compare NOA in different conditions.

3.4.3 Automated Assessment of Neurite Outgrowth Area

A detailed overview of all the steps in the automated analysis script is represented in Fig. 3. The analysis relies on three input values that determine the thresholding parameters. These parameters are obtained on the first picture of an entire experiment. All following explants will be automatically quantified using these input values. The setup of the analysis is described in the following steps.

1. Open the Axiovision software.
2. Use the "batch processing" function to select all the pictures of a single experiment. Select the custom made explant analysis script, as described in Gaublomme et al. [19]. Press start.

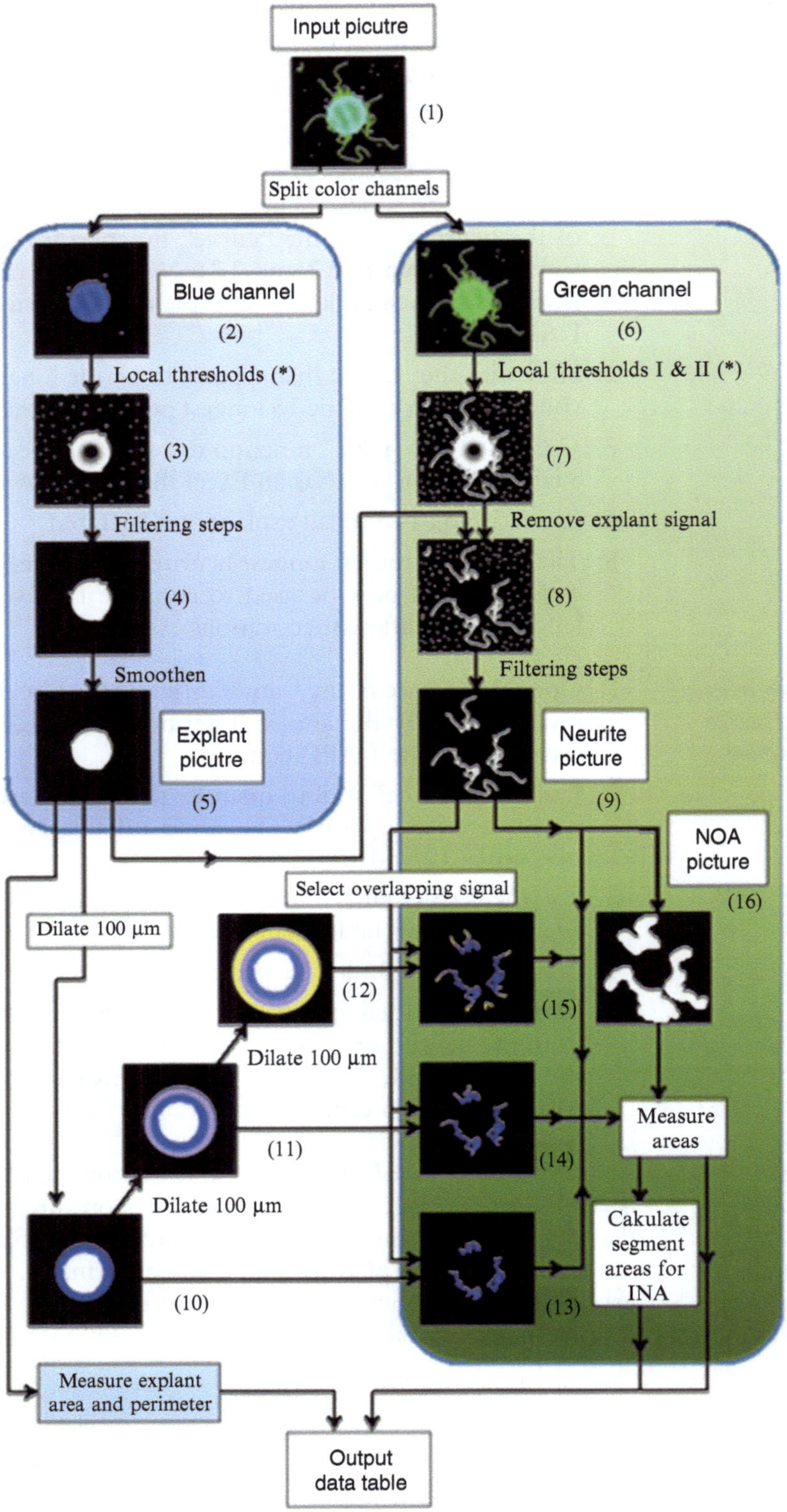

Input picutre
(1)
Split color channels
Blue channel
(2)
Local thresholds (*)
(3)
Filtering steps
(4)
Smoothen
Explant picutre
(5)
Green channel
(6)
Local thresholds I & II (*)
(7)
Remove explant signal
(8)
Filtering steps
Neurite picture
(9)
NOA picture
(16)
Select overlapping signal
Dilate 100 µm
(12)
(15)
Dilate 100 µm
(11)
(14)
Measure areas
Dilate 100 µm
(10)
(13)
Cakulate segment areas for INA
Measure explant area and perimeter
Output data table

3. Set the local threshold offset and size for the nuclear staining in the dialog screen (*see* **Note 14**) (Fig. 3, **step 3**) and press OK.
4. Set the threshold values for the neurite signal. Set the local threshold offset and size in the first dialog screen so that thin neurites are selected (*see* **Note 15**). Press OK. Set the local threshold offset and size in the second dialog screen so that neurites in dense areas are detected (Fig 3, **step 6**) (*see* **Note 16**). Press OK.
5. The analysis script will now automatically analyze all selected pictures from the experiment, as illustrated in Fig. 3. After thresholding a series of filtering steps is applied to generate the explant and neurite picture (Fig. 3, **steps 5** and **9**). Categorization of the INA is obtained by enlarging the explant body three times with 100 μm increments (Fig. 3, **step 10–12**) and selecting the overlapping signal with the neurite picture

Fig. 3 Schematic representation of "immunodetected neurite area" script. In the script measuring the "immunodetected neurite area," three input values, determining local thresholding parameters, are required. These values are determined on the first picture of a set of photographs from an entire experiment (at steps noted with *asterisk*). Once these parameters are set, all other pictures from the same experiment can be analyzed automatically, using the sequence of steps described below. Each input picture (*1*) (saved as a .tiff file) is split into separate color channels. The *blue* channel (*2*) visualizes the DAPI signal and is used for explant detection. The image is binarized with local thresholding (*3*). Next, in a series of subsequent filtering steps, all objects are dilated to close any gaps in the explant edge and afterwards filled up. Nuclei or tissue fragments that are slightly in contact with the explant are automatically separated and removed, together with any noise produced by local thresholding (*4*). Finally, the explant is slightly enlarged and smoothened and an explant picture (*5*) is generated, from which the explant area and perimeter are measured. The *green* channel (*6*) visualizes the β-tubulin signal and is used for neurite detection. Two local thresholds (*I* and *II*) are performed in parallel, differing in the size of the local area. Most neurites are detected using a threshold with a small local area (*I*). However, this setting may generate black gaps in dense neurite areas. To alleviate this problem, a second thresholding step with a larger local area (*II*) is used. Both output pictures are then combined in order to detect the entire neurite network (*7*). Next, the signal from the explant body itself is removed (*8*), followed by two subsequent filtering steps eliminating any remaining noise and artifacts. The end result is a binarized image of the undilated neurites without artifacts, the neurite picture (*9*). To categorize the total neurite area in different segments, the explant picture is dilated by 100 μm three times (*10–12*). Then, the overlapping signal between the neurite picture (*9*) and the dilated explant pictures (*10–12*) is selected, yielding pictures with neurites extending from the explant to the edge of various segments (*13–15*). Finally, from these resulting pictures (*13–15*) and from the total neurite picture (*9*) the neurite area is measured. From these measures, the immunodetected areas per segment are calculated by subtraction. The automated NOA measurement of the neurite outgrowth starts from the previously generated neurite picture. The detected neurites are dilated and all remaining holes in the neurite network are filled (*16*). From this NOA picture the neurite outgrowth area is measured. All values from one input picture are put, together with the image name, in a single row in an output data table, which will ultimately contain the data from all input pictures from one experiment after completion. The script also saves a jpeg file of the neurite picture (*9*) and the NOA picture (*16*) to quickly perform a quality control. Figure modified from ref. [19]

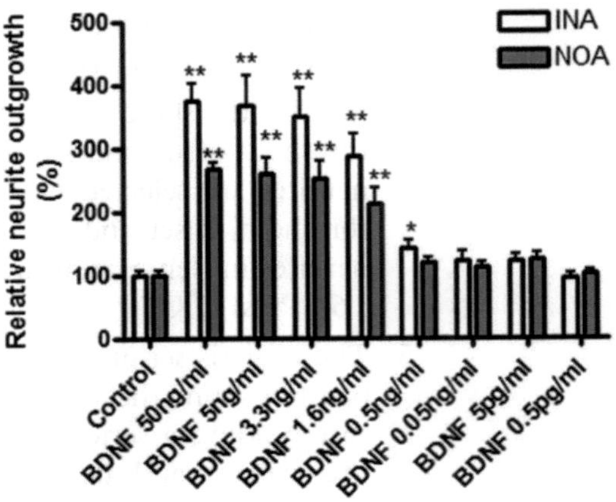

Fig. 4 Neurite outgrowth in retinal explants is stimulated after incubation with BDNF and CNTF. The neurotrophic factors BDNF and CNTF, in a 5:1 ratio, were added to the culture medium and explants were cultured for 3 days. Neurite outgrowth was plotted against the concentration of BDNF (the concentration of CNTF was adjusted accordingly). Both automated morphometric analyses, INA and NOA, showed a clear and dose-dependent stimulation of neurite outgrowth. Maximal stimulation was obtained at a concentration of 50 ng/mL BDNF and 10 ng/mL CNTF. Concentrations lower than 0.5 ng/mL BDNF and 0.1 ng/mL CNTF did not result in a significant stimulation of neurite outgrowth (*asterisk*) = $P<0.05$, (*double asterisk*) = $P<0.01$

(Fig. 3, **step 13–15**). Finally, the INA from all these pictures is measured and the INA per segment is calculated by subtraction.

6. For the quantification of the NOA, the detected neurites from the neurite picture are dilated. Next, all remaining holes in the neurite network are filled to generate the NOA picture (Fig. 3, **step 16**) from which the NOA is measured.
7. The automated analysis generates 2 jpeg files for each input picture, the neurite picture and the NOA picture. A single table, containing all measurements, is automatically saved as a csv file, which can be opened with Excel. For each explant a single row lists the following data: explant area and perimeter, the total and segmented INA values, and the NOA value.

3.5 Proof of Principle Experiment

As an example, a dose–response experiment using a combined administration of brain-derived neurotrophic factor and cilliary neurotrophic factor was performed (Fig. 4). Both INA and NOA show a clear and dose-dependent growth factor-induced upregulation of neurite outgrowth, as is described in literature. The combination of both neurotrophic factors can therefore be used a positive control in follow-up experiments.

4 Notes

1. Fresh explant culture medium is prepared for each experiment. Supplemented culture medium should not be stored for more than 7 days to ensure consistent results.
2. The primary and secondary antibodies listed provide excellent results for the immunological staining. There are other commercial anti-β-tubulin and secondary antibodies that give similar results.
3. Poly-L-lysine-coated plates can be stored in PBS for up to 2 days at 4 °C. Prolonged storage may result in reduced cell adhesion.
4. Depending on the protocol approved by the institute, mice can be euthanized by other methods.
5. The use of cold dissection medium will facilitate the dissection, as the retina is less prone to curling. Renew the dissection medium after every eye dissection to minimize the contamination with pigment cells.
6. Applying too much pressure will damage the bottom of the petri dish and cause plastic chips to attach to the explant, thereby interfering with the growth and the detection of the neurites. Make sure you excise the punches consistently at the same peripheral level. A large variation in the centro-peripheral position will introduce variability into the model.
7. Depending on the size of the punch needle and the age of dissection, 4–8 punches can be obtained from 1 retina. Using punch needles with a diameter of 750 μm, 6 punches can easily be harvested from a P0 retina and 8 punches from a P3 retina.
8. The RGC layer is located at the concave side of the explant. Furthermore, it has a clear crystal-like appearance. The opposite side of the retina frequently still has some dark-colored retinal pigment epithelium cells attached.
9. Depending on the type of experiment and age of mice, incubation can last for 2–7 days. Neurite outgrowth will be visible from 2 days on.
10. Incubation with the blocking solution, primary and secondary antibodies, and DAPI solution should be carried out in a dark and humid chamber to prevent evaporation of the buffers as well as photobleaching of the fluorophores.
11. Other forms of microscopy, such as phase contrast and DIC, may be used. Of note, glial cells also produce processes that extent past the explant body. Neurites can be identified as thin processes growing out of the explant body, whereas glial

processes have a thicker morphology. In our opinion superior results are obtained using immunohistochemical stainings for β-tubulin.

12. Alternatively, NIH image J software can be used. Different plugins, such as simple neurite tracer or NeuronJ, can assist in the tracing of the neurites.
13. Manual analysis should be performed blind to the experimental condition and by an experienced researcher in order to minimize user-biased variation.
14. The threshold value should be set so that the selected signal forms a closed circle, if not, the explant body cannot be detected.
15. The threshold value should be set so that the selected signal forms a continuous line. Individual pixels should not touch the detected neurite as this will result in a faulty measurement. In dense areas the neurite signal may not be detected at this stage, these denser areas will be detected in the following step. Afterwards, individual pixels will be filtered out (Fig. 3, **step 8**).
16. The threshold value should be set so that the neurite signal is detected in dense areas that are missed in the previous step. Thin neurites, detected in the previous step, may not be detected in this step.

Acknowledgments

Djoere Gaublomme is a fellow of the Research Foundation Flanders (FWO), Belgium, and Tom Buyens, Inge Van Hove and Lies De Groef are fellows of the Flemish Institute for the promotion of scientific research (IWT), Belgium. This work was supported by national grants from the Research Council of KU Leuven [KU Leuven BOF-OT/10/033] and the Research Foundation Flanders (FWO) [FWO G05311.10].

References

1. Berkelaar M, Clarke DB, Wang YC et al (1994) Axotomy results in delayed death and apoptosis of retinal ganglion cells in adult rats. J Neurosci 14(7):4368–4374
2. Fischer D, Leibinger M (2012) Promoting optic nerve regeneration. Prog Retin Eye Res 31(6):688–701
3. Goldberg JL, Klassen MP, Hua Y et al (2002) Amacrine-signaled loss of intrinsic axon growth ability by retinal ganglion cells. Science 296(5574):1860–1864
4. Kim J, Schafer J, Ming GL (2006) New directions in neuroregeneration. Expert Opin Biol Ther 6(8):735–738

5. de Lima S, Koriyama Y, Kurimoto T et al (2012) Full-length axon regeneration in the adult mouse optic nerve and partial recovery of simple visual behaviors. Proc Natl Acad Sci U S A 109(23):9149–9154
6. Bull ND, Johnson TV, Welsapar G et al (2011) Use of an adult rat retinal explant model for screening of potential retinal ganglion cell neuroprotective therapies. Invest Ophthalmol Vis Sci 52(6):3309–3320
7. Lin YP, Ouchi Y, Satoh S et al (2009) Sox2 plays a role in the induction of amacrine and Muller glial cells in mouse retinal progenitor cells. Invest Ophthalmol Vis Sci 50(1):68–74
8. Wang SW, Mu X, Bowers WJ et al (2002) Retinal ganglion cell differentiation in cultured mouse retinal explants. Methods 28(4): 448–456
9. Manabe S, Kashii S, Honda Y et al (2002) Quantification of axotomized ganglion cell death by explant culture of the rat retina. Neurosci Lett 334(1):33–36
10. Xin H, Yannazzo JA, Duncan RS et al (2007) A novel organotypic culture model of the postnatal mouse retina allows the study of glutamate-mediated excitotoxicity. J Neurosci Methods 159(1):35–42
11. Lagreze WA, Pielen A, Steingart R et al (2005) The peptides ADNF-9 and NAP increase survival and neurite outgrowth of rat retinal ganglion cells in vitro. Invest Ophthalmol Vis Sci 46(3):933–938
12. Toops KA, Berlinicke C, Zack DJ et al (2012) Hydrocortisone stimulates neurite outgrowth from mouse retinal explants by modulating macroglial activity. Invest Ophthalmol Vis Sci 53(4):2046–2061
13. Gasparini L, Crowther RA, Martin KR et al (2011) Tau inclusions in retinal ganglion cells of human P301S tau transgenic mice: effects on axonal viability. Neurobiol Aging 32(3): 419–433
14. Wigley CB, Berry M (1988) Regeneration of adult rat retinal ganglion cell processes in monolayer culture: comparisons between cultures of adult and neonatal neurons. Brain Res 470(1):85–98
15. Feng G, Mellor RH, Bernstein M et al (2000) Imaging neuronal subsets in transgenic mice expressing multiple spectral variants of GFP. Neuron 28(1):41–51
16. Schlamp CL, Johnson EC, Li Y et al (2001) Changes in Thy1 gene expression associated with damaged retinal ganglion cells. Mol Vis 7:192–201
17. Huang W, Fileta J, Guo Y et al (2006) Downregulation of Thy1 in retinal ganglion cells in experimental glaucoma. Curr Eye Res 31(3):265–271
18. Cregg JM, Wiseman SL, Pietrzak-Goetze NM et al (2010) A rapid, quantitative method for assessing axonal extension on biomaterial platforms. Tissue Eng Part C Methods 16(2): 167–172
19. Gaublomme D, Buyens T, Moons L (2013) Automated analysis of neurite outgrowth in mouse retinal explants. J Biomol Screen 18(5): 534–543

Chapter 6

Growth Cone Collapse Assay

Geoffrey M.W. Cook, Prem Jareonsettasin, and Roger J. Keynes

Abstract

The growth cone collapse assay has proved invaluable in detecting and purifying axonal repellents. Glycoproteins/proteins present in detergent extracts of biological tissues are incorporated into liposomes, added to growth cones in culture and changes in morphology are then assessed. Alternatively purified or recombinant molecules in aqueous solution may be added directly to the cultures. In both cases after a defined period of time (up to 1 h), the cultures are fixed and then assessed by inverted phase contrast microscopy for the percentage of growth cones showing a collapsed profile with loss of flattened morphology, filopodia, and lamellipodia.

Key words Axon growth repellents, Growth cones, Dorsal root ganglia, Liposomes, Phase contrast microscopy

1 Introduction

The growth cone collapse assay has become a standard method for assaying and identifying molecules that inhibit neuronal growth cone motility. Without this simple bioassay, the biochemical purification of neuronal repellents would have been considerably delayed. The assay was first described by Raper and Kapfhammer in 1990 [1], and the roots of their method lie in earlier observations on the formation of neurite territories when dissimilar explants are adjacent to one another [2, 3].

Using time-lapse video microscopy, Kapfhammer and Raper [4] studied what happened when growth cone-neurite encounters take place in vitro. They found that when retinal growth cones contact retinal neurites, or sympathetic growth cones contact sympathetic neurites, the growth cones retain their normal, flattened hand-like morphology with lamellipodia and filopodia. However, when retinal growth cones make contact with sympathetic neurites, or sympathetic growth cones contact retinal neurites, growth cones undergo an abrupt loss of their hand-like appearance—growth cone collapse—and cease advancing (Fig. 1).

Andrew J. Murray (ed.), *Axon Growth and Regeneration: Methods and Protocols*, Methods in Molecular Biology, vol. 1162,
DOI 10.1007/978-1-4939-0777-9_6, © Springer Science+Business Media New York 2014

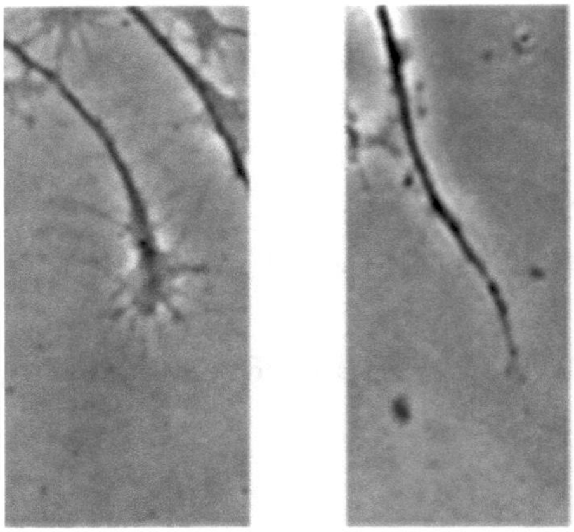

Fig. 1 *Left panel*, typical spread morphology of a chick DRG axon growth cone extending on poly-L-lysine/laminin substrate, showing flattened lamellipodia, and protruding filopodia. *Right panel*, growth cone collapse after addition to culture of liposomes derived from detergent extract of chick somite tissue; magnification 500×

Raper and Kapfhammer [1] reasoned that such specific motility-inhibiting effects operating in vivo could determine which of several alternative pathways a growth cone will take. Their working hypothesis was that neurites have membrane-associated labels that initiate the collapse of those growth cones that are able to detect them. The assay they devised was intended to allow identification of these labels at the molecular level. Suspensions of test membranes, or detergent-solubilized membranes from which the detergent has been removed, are added to cultures of chick dorsal root ganglia (DRG) extending axons on a permissive substrate. After a defined period the cultures are fixed and scored for the percentage of growth cones that exhibit a collapsed morphology. The original study [1] confirmed the hypothesis that specific membrane-associated signals can shut down growth cone motility in an environment otherwise permissive for outgrowth. Importantly it also showed that the assay can be used as a means of identifying biochemically enriched fractions of growth cone collapse-inducing activity, so providing a convenient method to monitor and modify purification protocols for the identification of biologically active growth repellents and their specificities.

In the late 1980s our laboratory began a series of experiments in the chick embryo aimed at establishing whether the segmented pattern of spinal nerves might be generated by inhibitory/repulsive cues expressed by sclerotome cells of the posterior halves of the somites [5]. At that time we had devised a simple bioassay based on measuring the extent of axon growth from chick DRGs cultured on nitrocellulose substrates to which test molecules had been adsorbed. In this assay the radius of axon growth was related only to the substrate and not to the size of the explanted pieces of ganglia. The assay allowed us to show that glycoproteins derived from the posterior half-sclerotome inhibit axon growth, suggesting that they could be components of a growth cone guidance system in vivo, but the technique is not without limitations. Individual neurites cannot be distinguished, and to mitigate this only those axons that migrate over the longest distance are measured, and these axons may be atypical in their response. In addition the lack of resolution of individual axons makes it impossible to decide whether axons are advancing as a direct result of the nature of the substrate or because they are advancing along other pioneer axons. At this time Dr Raper, then working in Tübingen, Germany, visited our laboratory, saw our results and told us of his developing growth cone collapse assay. Following a kind invitation from Drs Kapfhammer and Raper, Jamie Davies (at that time a graduate student in our laboratory) was invited to visit Tübingen to learn their methodology. As a result we were able to use the assay to show growth cone collapse-inducing activity in chick embryo somites [5].

As pointed out by Raper and colleagues [1, 6] growth cone collapse serves as a very sensitive and reliable indicator of repellent activity, and can be used to titrate the concentration dependence of repellent activity. The assay does not require any specialized equipment and can be set up in any laboratory with experience of neuronal cell culture. Growth cone morphology is preserved by fixation and the cultures can be assed for percentage collapse using phase contrast microscopy at any time convenient to the experimenter (*see* Fig. 2 for summary).

We should also point out the potential utility of the assay in high-throughput screening for small molecules that modulate growth cone collapse. This would necessitate the use of appropriate image software to assess the degree of collapse. For example, several mood-stabilizing drugs in current clinical use have been shown to inhibit spontaneous collapse of neonatal rat DRG growth cones in vitro [7], and this phenomenon has been investigated at GlaxoSmithKline using such software in the context of potential drug treatments for bipolar disorder [8].

In this chapter we describe the method used in our laboratory which can be readily accomplished within 2–3 days.

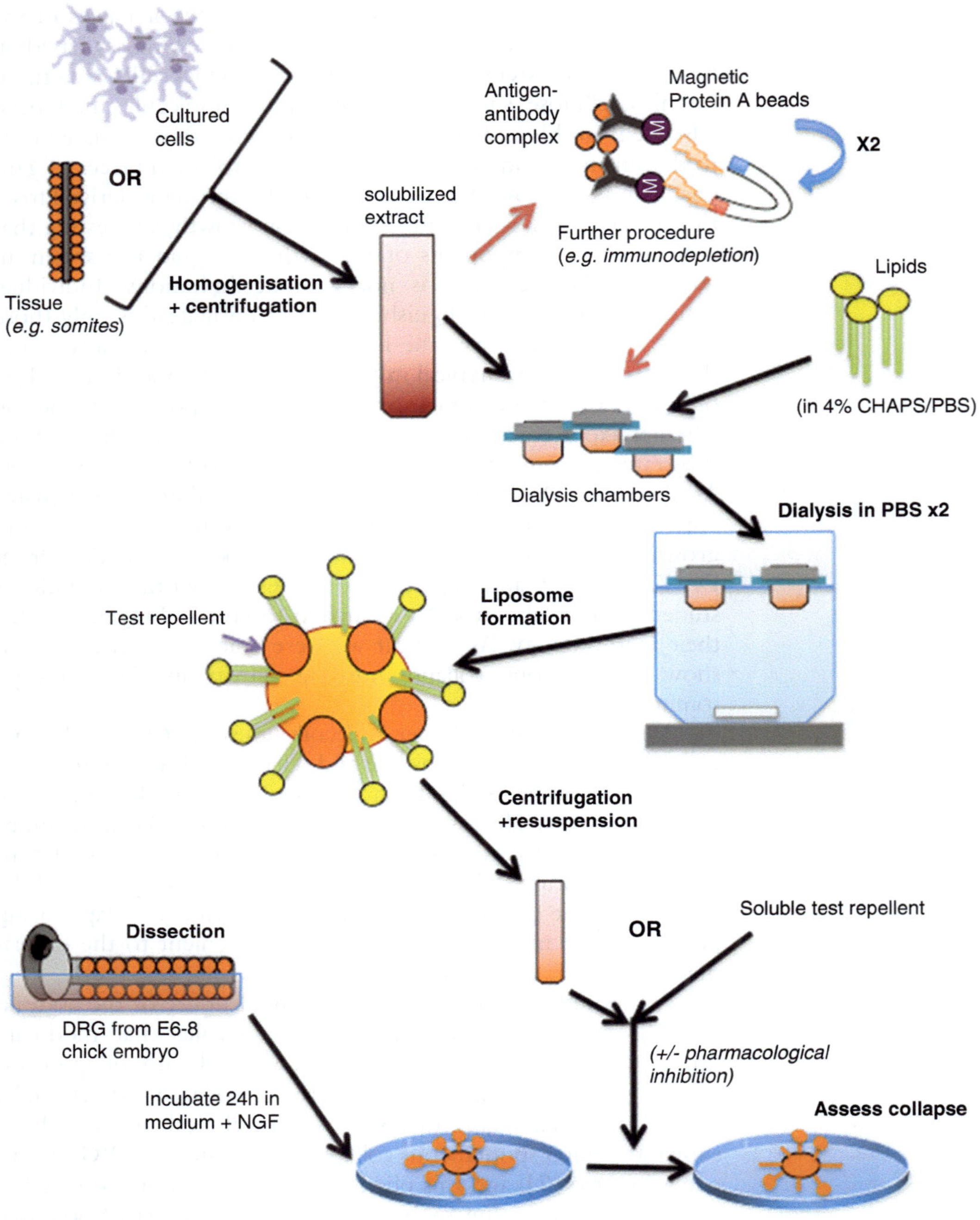

Fig. 2 Schematic diagram showing the main steps of the growth cone assay described in the text, starting (*top left*) with tissue/cell extraction, and ending (*bottom right*) with the addition of liposomes or soluble repellent to axon growth cones

2 Materials

2.1 Chick Dorsal Root Ganglion Culture and Fixation

Prepare all solutions using ultrapure water (18 MΩ cm deionized water at 25 °C is ideal). Reagents should be of analytical grade and solutions are made up fresh before each assay, with the exception of fixative which can be stored at 4 °C in a dedicated refrigerator. The disposal of reagents should accord with the local regulations.

1. Chick embryo (*Gallus gallus*): Incubate at 37 °C to gestational age 6–8 days, Hamburger and Hamilton stages 29–34 [9] (*see* **Note 1**).
2. 13 mm Diameter glass coverslips (VWR, catalog number 631–0149): Clean in concentrated nitric acid (30 min; *see* **Note 2**), wash repeatedly in purified water as determined by monitoring the pH to confirm the removal of all the acid, and then wash 5× in 99.9 % ethanol; store in ethanol in a sealed vessel at room temperature.
3. Spirit burner on a metal tray.
4. Dulbecco's modified Eagle's medium (DMEM, Sigma).
5. Poly-L-lysine (0.01 %, Sigma, catalog number P4707).
6. Laminin (Sigma, catalog number L2020): Use at 40 μg/ml in DMEM.
7. DRG culture medium: DMEM with 2 % B-27 supplement (Invitrogen, catalog number 17504044), 1 % L-glutamine–penicillin–streptomycin solution (Sigma, catalog number G1146) and 40 ng/ml murine NGF-7S (Sigma, catalog number N0513-1MG; *see* **Note 3**).
8. Petri dishes: 35, 50, and 100 mm diameter.
9. Nunc 4-well Multidish, 66 mm × 66 mm.
10. Dissection equipment: Micro-feather ophthalmic scalpel, John Weiss International; dissection scissors; watchmaker's forceps; Sylgard (Corning)-coated dishes; sterile dissection pins.
11. Gilson pipette, P20.
12. Dissecting microscope.
13. Inverted phase-contrast microscope.
14. Fixative solution: 4 % (w/v) formaldehyde in 15 % sucrose. This is prepared by mixing paraformaldehyde (40 g) with 10× PBS (100 ml) and analytical grade water (600 ml) on a heated magnetic stirrer in a fume cupboard. 5–6 drops of 10 M NaOH are then added whilst heating to 55–65 °C, and the mixture stirred until the paraformaldehyde has dissolved (1–2 h). The heater is then turned off and sucrose (100 g) is

added with stirring until dissolved. The solution is then allowed to cool at room temperature, its pH adjusted to 7.4, and then made up to 1 l with water in a volumetric flask, filtered (in the fume cupboard) through Whatman no.1 paper. The clear filtrate is stored at 4 °C.

2.2 Growth Repellent Molecules

1. Test repellent: crude extract, purified protein, or recombinant protein.
2. 2 % CHAPS in PBS with protease inhibitor cocktail: CHAPS, 3-[(3-cholamidopropyl)dimethylammonio]-1-propanesulfonate, is a zwitterionic detergent that is particularly useful for membrane work being a very good solubilizer and non-denaturing. As a zwitterionic detergent it can be used for isoelectric focusing or native polyacrylamide gel electrophoresis. Because of its high critical micellar concentration, it is easily removed from protein/detergent solutions by dialysis. It is suitable for membrane reconstitution studies or for the formation of artificial membranes (liposomes) as used in this assay. We obtain CHAPS from ICN Biomedicals, Inc., Ohio (there are many other suppliers of this detergent). Protease inhibitor cocktail from Roche, "complete" tablets in EASYpack, which inhibits a broad spectrum of serine, cysteine, and metalloproteases as well as calpains (other suppliers could be used). We find it convenient to use PBS (Dulbecco A) tablets supplied by Oxoid.
3. Glass Dounce homogenizer or a grinding kit (GE Health Care, catalog number 80-6483-37).
4. Ultracentrifuge.
5. Phospholipids: Add 1,2-diacyl-sn-glycero-3-phospho-L-serine (Sigma, catalog number P6641; 10.2 mg/ml in chloroform/methanol) and L-α-phosphatidylcholine (Sigma, catalog number P2772; 100 mg/ml in chloroform) in a ratio of 0.98:1 to a small glass tube (*see* **Note 4**). Dry in fume cupboard over night to allow solvent to evaporate. Dissolve in 4 % CHAPS in PBS; the dried lipids should be resuspended with constant mixing using a microspatula until clear.
6. Dialysis chambers: Construct small dialysis chambers from the lids of Eppendorf plastic microcentrifuge tubes. As a sealing ring for the chamber use the ringed top of the Eppendorf tube, cut from the remainder of the tube with a razor blade. Boil chambers and rings 3× in purified water.
7. Dialysis membrane (Medicell International, Visking size 2–18/32 in., MW cut-off 12–14 kDa): boil and wash 5× using in purified water, cut into squares (~2 cm × 2 cm), store in purified water in a sealed vessel.

3 Methods

All procedures are carried out at room temperature unless otherwise stated.

3.1 Preparation of Coverslips for DRGs (Day 1 of Experiment)

1. Flame-dry acid-cleaned and ethanol-washed coverslips to remove ethanol by passing through the flame of a spirit burner on a metal tray. Place coverslips around the circumference of a 100 mm diameter Petri dish. Coat coverslips in poly-L-lysine on one side by creating coverslip "sandwiches": place 30 μl 0.01 % poly-L-lysine on one coverslip and cover this with a second coverslip. Incubate sandwiches for 1 h at 37 °C (*see* **Note 5**).
2. Wash coverslips 3× in DMEM to remove excess poly-L-lysine. Reform sandwiches with 40 μl laminin (40 μg/ml in DMEM) and incubate for 1 h at 37 °C (*see* **Note 6**).
3. Prepare 4-well Nunc dishes with 500 μl culture medium per well.
   - Wash coverslips 1× in DMEM to remove excess laminin. Place one poly-L-lysine/laminin-coated coverslip in each well of 4-well Nunc dish. Ensure that the coated side of each coverslip faces upwards and that the coverslip is placed at the bottom of the well. Incubate in 5%CO_2 at 37 °C for ~24 h until further use.

3.2 Preparation of Test Repellent (Day 1 of Experiment)

Soluble repellents may be added directly to the culture; non-soluble repellents are first incorporated into liposomes to form a suspension. Tissues or cell cultures being examined for repellent activity are homogenized prior to incorporation into liposomes.

1. Homogenize tissue- or cell culture-derived material in 2 %CHAPS in PBS/protease inhibitor cocktail using glass Dounce homogenizer or grinding kit.
2. Ultracentrifuge homogenate at 100,000×*g* av. for 1 h at 4 °C. Retain the supernatant fluid and dilute as appropriate. Perform further manipulations (e.g., immunodepletion) after this step as required.
3. Add solubilized extract (which may be further treated as **step 2** above) and phospholipids (solubilized as Subheading 2.2, **item 5** above) to dialysis chambers. In each 100 μl chamber mix 10 μl phospholipids with 90 μl solubilized extract. Cover the top of each dialysis chamber with a 2 cm×2 cm dialysis membrane (the membrane can be removed from storage with forceps and the excess water removed by gentle blotting on tissue paper), place the ring horizontally on top of the

membrane and seal the chamber by pressing the ring firmly and evenly down and around the sides of chamber, with gloved hands. Confirm that the membrane is not split during this procedure by gently pressing the membrane with blunt forceps after the procedure.

4. Float the chambers membrane-down in 2 l PBS and dialyze twice, each for 12 h at 4 °C, with gentle stirring on a magnetic stirrer. The detergent dialyzes out, resulting in the formation of liposomes incorporating the test repellent. After dialysis, on day 2 of the experiment, the floating chambers are removed and excess dialyzate adhering to the chambers removed by blotting with tissue paper. The contents of each chamber are removed by puncturing the membrane with the tip of a Gilson pipette and the contents withdrawn. It is recommended that gloved hands are used in this procedure.
5. Centrifuge the liposome suspension at 100,000 × *g* av. for 1 h at 4 °C and resuspend the liposome pellet in PBS or, if appropriate, in a solution of putative inhibitor of the test repellent.

3.3 Dissection of Chick Embryo DRGs (Day 1 of Experiment)

1. Sterilize and clean all microdissection equipment using 70 % ethanol prior to Dissection.
2. Fill two 35 mm Petri dishes with cold DMEM (4 °C). Fill one 35 mm and one 50 mm Sylgard-coated dish with cold culture medium.
3. Remove embryo from the egg into a dry 100 mm Petri dish and decapitate using dissecting scissors. Transfer the embryo to the 35 mm Petri dish filled with cold DMEM and wash thoroughly to remove any adherent yolk and albumen. Transfer the embryo to the 50 mm Sylgard-coated dish.
4. Under the dissecting microscope, pin the embryo on its right-hand side so that the ventral thoraco-abdominal organs are placed to the left and the dorsal vertebral column/neural tube to the right. Cut away the organs using a microscalpel (*see* **Note** 7).
5. Pin the vertebral column/neural tube (together with overlying ectoderm) dorsal-up. Strip away the ectoderm and bisect the column along the dorsal midline using the microscalpel.
6. From each half vertebral column strip away the meninges and neural tube using watchmaker's forceps to reveal a segmented series of DRGs with intervening vertebral pedicles. Remove each DRG using watchmaker's forceps and transfer DRGs en-masse into the 35 mm Sylgard-coated dish with culture medium using a P20 Gilson pipette (*see* **Note 8**).

7. Using a microscalpel, trim away adherent tissue and bisect each DRG.
8. Transfer DRGs using a P20 Gilson pipette to a new 35 mm Petri dish filled with cold DMEM to remove remaining tissue debris.
9. Using the P20 Gilson pipette place 1–3 DRG explants onto each coverslip in the 4-well dishes. Ensure explants are well separated from each other.
10. Incubate in 5 %CO_2 for ~24 h at 37 °C.
11. Examine the cultures using an inverted phase-contrast microscope. Select for assay those showing substantial axon outgrowth, spread growth cone morphology (approximately no more than 20 % should appear collapsed), and lack of non-neuronal cells.

3.4 Growth Cone Collapse Assay (Day 2 of Experiment)

1. Prepare test axon growth repellent as required prior to assay.
2. Blind-code and randomly assign repellent additives for each culture in order to minimize observer bias.
3. Slowly pipette solubilized or liposome-formed test repellents directly over the DRGs, ensuring that repellent is thoroughly mixed in each well.
4. Incubate dishes in 5% CO_2 at 37 °C for 30 min to 1 h, depending on dynamics of collapse induced by repellent under test.
5. Fix cultures by removing 250 μl medium slowly from the side of each well, followed by slow addition of 250 μl of fixative solution, also at the side so as not to dislodge explants mechanically. The fixative solution is denser than the culture medium and initially settles below the medium. Repeat this step immediately, aspirating from the upper part of the fluid, to replace as much medium as possible with fixative. Leave cultures at room temperature for 1 h and store at 4 °C in PBS prior to scoring.
6. View with an inverted phase contrast microscope using the 40× objective (*see* **Note 9**).
7. Examine growth cones extending at the outermost perimeter of the explant, scoring those that do not make contact with other axons. Cultures suitable for scoring should contain at least 100 growth cones, of which no more than 20 % should be collapsed under control conditions. Score each growth cone as either *spread* with 2 or more filopodia and/or flattened lamellipodia, or *collapsed* with a bullet-shaped morphology or 2 or less filopodia and no lamellipodia. Express the result as a percentage of growth cone collapse for each DRG.

4 Notes

1. One chick embryo yields ~10–20 DRGs.
2. Considerable caution should be used when handling nitric acid. Carry out all procedures in a well-ventilated fume hood. Personal protection is essential—laboratory coat and apron, gloves, and safety goggles. A convenient method is to cover the coverslips in a beaker with nitric acid also containing several small glass anti-bumping balls [6]. The beaker is covered with a watchglass and heated slowly towards boiling temperature on a hot plate for between 15 and 30 min, after which the acid is poured carefully into an acid waste receptacle and the coverslips are washed in five changes of deionized water as described above.
3. The NGF concentration can be altered between 5 and 80 ng/ml to ensure good axon outgrowth and spread growth cones.
4. 10 μl lipids is sufficient for 90 μl test repellent in a single dialysis chamber.
5. A 100 mm diameter Petri dish accommodates up to 12 coverslip sandwiches.
6. Separate each coverslip sandwich using two pairs of watchmaker's forceps: move apart the top and bottom coverslips; hold the bottom coverslip down and pick up the top cover slip; remove the bottom coverslip from the dish by pushing it against the edge of the dish and lifting it up at the other side.
7. To facilitate dissection, ensure the embryo is stretched slightly along the anterior–posterior axis and pin it down at the neck and between the left hip joint and tail. To remove the organs start the cut at the level of the left shoulder joint and extend anteriorly and posteriorly.
8. Each DRG lies in a cavity between adjacent vertebral pedicles. To remove the DRG expand the cavity with watchmaker's forceps and avulse the DRG from its peripheral spinal nerve root with a second pair of forceps.
9. Remove each coverslip from the 4-well dish using the bent tip of a hypodermic needle to lift up the coverslip against the edge of the dish so it can then be gripped with watchmaker's forceps. Place the coverslip with DRG upwards in a 35 mm Petri dish filled with PBS, press the coverslip down on the base of the dish with forceps to remove intervening fluid and so stabilize the coverslip for phase contrast microscopy. As an alternative to viewing by phase contrast, stain growth cones at this stage for cytoskeletal proteins if preferred.

References

1. Raper JA, Kapfhammer JP (1990) The enrichment of a neuronal growth cone collapsing activity from embryonic chick brain. Neuron 4: 21–29
2. Kapfhammer JP, Grunewald BE, Raper JA (1986) The selective inhibition of growth cone extension by specific neurites in culture. J Neurosci 6:2527–2534
3. Kapfhammer JP, Raper JA (1987) Interactions between growth cones and neurites growing from different neural tissues in culture. J Neurosci 7:1595–1600
4. Kapfhammer JP, Raper JA (1987) Collapse of growth structure on contact with specific neurites in culture. J Neurosci 7:201–212
5. Davies JA, Cook GM, Stern CD, Keynes RJ (1990) Isolation from chick somites of a glycoprotein fraction that causes collapse of dorsal root ganglion growth cones. Neuron 4: 11–20
6. Kapfhammer JP, Xu H, Raper JA (2007) The detection and quantification of growth cone collapsing activities. Nat Protoc 2:2005–2011
7. Williams RS, Cheng L, Mudge AW, Harwood AJ (2002) A common mechanism of action for three mood-stabilizing drugs. Nature 417: 292–295
8. Di Daniel E (2007) Growth cone collapse assay at GlaxoSmithKline. http://www.yumpu.com/en/document/view/4935410/growth-cone-collapse-assay-at-glaxosmithkline-definiens
9. Hamburger V, Hamilton HL (1951) A series of normal stages in the development of the chick embryo. J Morphol 88:49–92

Chapter 7

Axon Length Quantification Microfluidic Culture Platform for Growth and Regeneration Study

Jaewon Park, Sunja Kim, Jianrong Li, and Arum Han

Abstract

To fully understand how external biomolecular environment influences axon growth, a method that can easily quantify the extent of axon growth as well as locally control their biomolecular environment is critically needed. Here, we describe a microfluidic culture platform capable of isolating CNS axons from neuronal somata for localized biomolecular manipulation as well as providing linearly guided axon growths for simple and easy quantification of the axon growth length. The axon isolation and guidance capability combined with the multi-compartment configuration make this platform ideal for investigating and screening drugs or other molecular factors that promote axon growth as well as regeneration.

Key words Neuron culture microsystem, Compartmentalized culture, Axon isolation, Axon guidance, Axon growth quantification, Localized biomolecular treatment

1 Introduction

Adult mammalian central nervous system (CNS) cannot be functionally recovered after injury mainly due to the inability of the damaged CNS axons to regenerate. It has been thought that extrinsic environmental factors such as myelin-associated inhibitors are the dominant factor that prevent CNS axons from regenerating [1, 2]. However, it has been shown that manipulation of axon growth pathways can also promote CNS axon regeneration after injury [3]. This suggests that not only the environmental factors but also the growth capability of the neurons take important role in axon regeneration.

In order to better understand the growth or regeneration mechanism of axons and to find therapeutic treatments that enhance axon growth or regeneration, methods capable of providing controlled and localized biomolecular microenvironments while allowing easy axon growth quantification are necessary. However, conventional culture plate-based in vitro neuron culture methods are limited in providing such features. First, in most

Andrew J. Murray (ed.), *Axon Growth and Regeneration: Methods and Protocols*, Methods in Molecular Biology, vol. 1162, DOI 10.1007/978-1-4939-0777-9_7,

conventional culture methods other than the "Campenot chamber" [4], both neuronal somata and axons are exposed to the same biochemical environment, which hinders investigating localized effects of biomolecular treatments independently. Second, tracking and quantitatively analyzing the extent of axon growth in response to these factors over long culture period at commonly used in vitro culture density is challenging. Most of axon growth studies are often performed at a significantly lower cell density (3–20 cells/mm^2) and for only a short period of time after cell seeding (<5 days) [5–10]. This is largely due to difficulties in tracking and analyzing the length of randomly grown axons.

Several compartmentalized neuron culture microsystems capable of isolating axons from neuronal somata or dendrites have been previously introduced by us and others [11–13]. However, the growth direction of axons was not controlled and axons formed intensively tangled network upon isolation, making it almost impossible to track or quantitatively analyze individual axons. Therefore, an in vitro neuron culture platform that provides axonal isolation with a capability to easily quantify axonal growths could provide important advances in finding biochemical factors or pharmaceuticals that promote the growth capability of CNS axons.

In this chapter, we describe the fabrication and preparation steps for a neuron culture platform where axons are physically guided to grow in straight lines and can be fluidically and physically isolated from neuronal somata and dendrites into six satellite axon compartments for localized biomolecular treatments [14]. The neuron culture platform is composed of two poly (dimethylsiloxane) (PDMS) layers and a PDMS ridge structure (Fig. 1a). The top PDMS layer is composed of one soma compartment and six axon compartments that are 800 μm apart. The ridge structure that is assembled at the center of the soma compartment

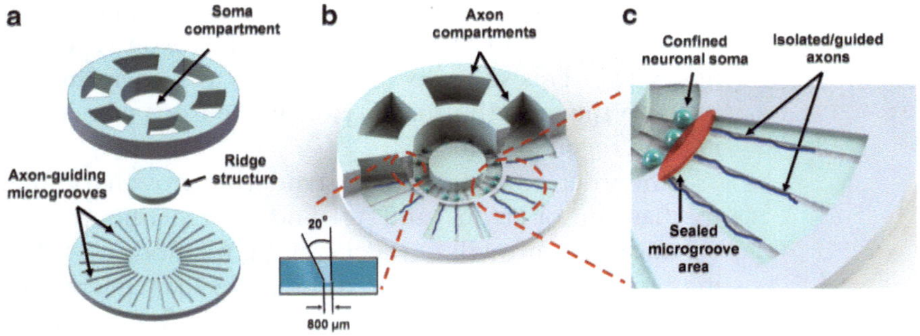

Fig. 1 (**a**) PDMS components of the developed quantitative axon growth analysis neuron culture platform. (**b**, **c**) Illustration showing axon isolation as well as axon guidance via the array of shallow microgrooves patterned on the bottom substrate. (**b**-*Inset*) Cross-sectional view showing 20° inclined PDMS wall separating the soma and the axon compartment (*see* **Note 4**)

reduces the number of required neurons by minimizing the effective cell culture area and increases the axon isolation efficiency. The array of shallow axon-guiding microgrooves ($3 \times 20 \times 10{,}000\ \mu m^3$) patterned on the bottom PDMS layer has two functions. First, it connects the soma compartment and the axon compartments. The microgrooves confine neuronal somata inside the soma compartment while allowing axons to pass through for isolation inside the six axon compartments (Fig. 1b). Second, the microgrooves guide axons to grow in straight lines for easy quantitative axonal length analysis (Fig. 1c). Therefore, quantitative analysis of the axon length in response to different localized biomolecular treatments can be performed.

The culture platform introduced in this chapter not only can be used for studying the axonal growth upon localized biomolecular treatments but can also be utilized for studying axonal transection and regeneration, axonal transport under tightly controlled biochemical environments, or studying differential effects of a molecule on somata and axons. Also, the design can be modified to have increased number of axon compartments for achieving higher throughput.

2 Materials

2.1 Equipment

CNC milling machine, digital balance, laboratory oven, mask aligner, spin-coater, hot plate, oxygen plasma chamber, autoclave, vacuum connected desiccator, humidified 5 % CO_2 incubator.

2.2 Materials

2.2.1 Culture Platform

- 3 in. Silicon wafer (SQI, Inc.).
- SU-8™ 2002 photoresist (Microchem).
- SU-8™ developer (propylene glycol methyl ether acetate (PGMEA), Microchem).
- Transparency film mask (CAD/Art service).
- Isopropyl alcohol (EMD Millipore).
- Sulfuric acid (EMD Millipore).
- Hydrogen peroxide (BDH Chemicals).
- (tridecafluoro-1,1,2,2-tetrahydrooctyl)-Trichlorosilane (United Chemical Technologies, Inc.).
- Poly(methyl methacrylate) (PMMA) block (McMaster-Carr).
- Poly(dimethylsiloxane) (PDMS) (Sylgard 184, Dow Corning).
- Sharpened stainless steel punch (Technical Innovations, LLC).
- Poly-D-lysine (PDL) (Sigma-Aldrich).
- Calcein-AM (Invitrogen).
- Weighing boat (VWR International).
- Disposable stir rod (e.g., plastic fork).

- Razor blades (VWR International).
- Parafilm (VWR International).
- Disposable petri dish (VWR International).

2.2.2 Neuron Culture

- Sterilized petri dish, forceps, and scissors.
- NBB27 with glutamate (DIV 0–DIV 4): Neurobasal medium (100 ml, Invitrogen), B27 supplement (2 ml, Invitrogen), L-glutamine (1 mM, Sigma-Aldrich), L-glutamic acid (25 μM, Sigma-Aldrich), *N*-acetyl cysteine (63 ng/ml, Sigma-Aldrich), penicillin/streptomycin (1 ml, Invitrogen).
- NBB27 (from DIV 4): Neurobasal medium (100 ml, Invitrogen), B27 supplement (2 ml, Invitrogen), GluMAX (0.5 mM, Invitrogen), *N*-acetyl cysteine (63 ng/ml, Sigma-Aldrich), penicillin/streptomycin (1 ml, Invitrogen).

3 Methods

3.1 Culture Platform Fabrication

3.1.1 Bottom Layer Fabrication (Fig. 2)

1. Clean a silicon wafer with Piranha solution (H_2O_2:H_2SO_4 = 1:3) at 180 °C for 20 min (*see* **Note 1**).
2. Thoroughly rinse Piranha cleaned wafer with DI water and dry it with pressurized nitrogen gas.
3. Spin-coat SU-8™ 2002 photoresist on the cleaned silicon wafer at 750 rpm for 30 s to obtain a film thickness of 3 μm.
4. Soft-bake the SU-8™ spin-coated wafer at 95 °C for 5 min on a leveled hot plate.
5. Expose the wafer through a transparency film mask (20,000 dpi) having the microgroove array and align mark patterns with the dosage of 105 mJ/cm^2.

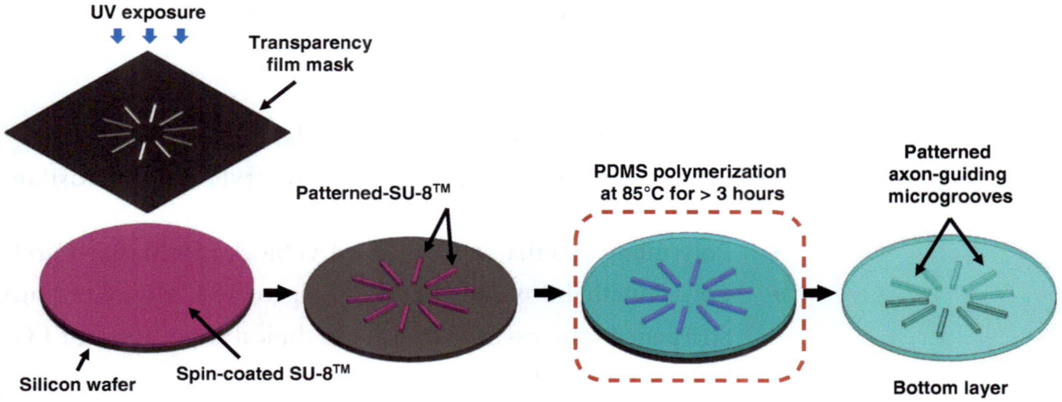

Fig. 2 Illustrations showing the fabrication process of the bottom PDMS layer by conventional photolithography and PDMS soft-lithography processes

6. Post-exposure bake the wafer at 95 °C for 5 min on a leveled hot plate.
7. Develop the wafer with SU-8™ developer (PGMEA) for 45–60 s to remove uncross-linked SU-8™.
8. Rinse the developed wafer with IPA and dry it with pressurized nitrogen gas (*see* **Note 2**).
9. Vapor-coat SU-8™ patterned silicon wafer with (tridecafluoro-1,1,2,2-tetrahydrooctyl)-trichlorosilane for 20 min inside a vacuum chamber, followed by a brief rinsing step with IPA (*see* **Note 3**).
10. Thoroughly mix PDMS prepolymer with curing agent (10:1 w/w) in a weighing boat using a stir rod (*see* **Note 5**).
11. Place the SU-8™ master mold in a disposable container and slowly pour the PDMS mixture until the PDMS reaches 1–2 mm above the master mold surface.
12. Remove excessive bubbles in the PDMS mixture by placing the PDMS-poured SU-8™ master mold in a vacuum chamber for 10–15 min.
13. Place the PDMS-poured SU-8™ master mold inside a leveled drying oven and cure at 85 °C for at least 3 h.
14. After PDMS has been fully polymerized, detach the bottom layer from the SU-8™ master mold (*see* **Note 6**).

3.1.2 Top Layer Fabrication (Fig. 3)

PMMA Master Mold Fabrication

1. Design one soma compartment and six satellite axon compartments that are 800 μm apart from the soma compartment using any computer-aided design (CAD) program (*see* **Note 4**). The distances can be adjusted based on the user's need.
2. Pattern the compartment structures on the poly(methyl methacrylate) (PMMA) block using a computer numerically controlled (CNC) milling machine.
3. Rinse the PMMA master mold with IPA three times.

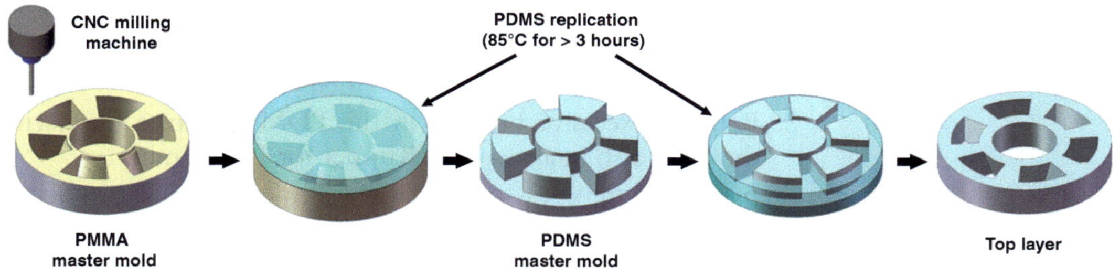

Fig. 3 Illustrations showing the fabrication process of the top PDMS layer by double PDMS soft-lithography process using CNC milling machine engraved PMMA as a master mold

PDMS Master Mold Fabrication

1. Thoroughly mix PDMS prepolymer with curing agent (10:1 w/w) in a weighing boat using a stir rod (*see* **Note 5**).
2. Place the PMMA master mold in a disposable container and slowly pour the PDMS mixture until the PDMS reaches 3–5 mm above the PMMA master mold surface.
3. Remove excessive bubbles in the PDMS mixture by placing the PDMS-poured PMMA master mold in a vacuum chamber for 10–15 min.
4. Place the PDMS-poured PMMA master mold inside a leveled drying oven and cure at 85 °C for at least 3 h.
5. After PDMS has been fully polymerized, detach the PDMS master mold from the PMMA master mold (*see* **Note 6**).
6. Vapor-coat the detached PDMS master mold with (tridecafluoro-1,1,2,2-tetrahydrooctyl)-trichlorosilane for 20 min inside a vacuum chamber, followed by a brief rinsing step with IPA.

Compartment Layer Fabrication

1. Thoroughly mix PDMS prepolymer with curing agent (10:1 w/w) in a weighing boat using a stir rod.
2. Remove excessive bubbles in the PDMS mixture by placing it in a vacuum chamber for 10–15 min (*see* **Note 7**).
3. Place the PDMS master mold in a disposable container and slowly pour the PDMS mixture until the PDMS reaches 2–3 mm below the top of the PDMS master surface.
4. Remove excessive bubbles in the PDMS mixture by placing the PDMS-poured PDMS master mold in a vacuum chamber for 10–15 min.
5. Place the PDMS-poured PDMS master mold inside a leveled drying oven and cure at 85 °C for at least 3 h.
6. Detach the final PDMS top layer from the PDMS master mold and cut out rough boundaries with a razor blade to make the bottom surface perfectly flat.

3.1.3 Ridge Structure Fabrication

1. Thoroughly mix PDMS prepolymer with curing agent (10:1 w/w) in a weighing boat using a stir rod.
2. In order to obtain a PDMS slab with approximately 2 mm thickness, pour 10 ml of PDMS mixture to a petri dish.
3. Remove excessive bubbles in the PDMS mixture by placing the petri dish in a vacuum chamber for 10–15 min.
4. Place the PDMS-poured petri dish inside a leveled drying oven and cure at 85 °C for at least 3 h.
5. Cut out a PDMS ridge structure using a round-hole punch.

3.2 Device Assembly and Neuron Culture Preparation Steps

1. Place a top layer, a ridge structure, and a bottom layer (surfaces to be bonded facing up) inside the oxygen plasma chamber.
2. Turn on vacuum of the plasma chamber for 2 min prior to turning on the plasma and perfuse oxygen gas for 2 min upon ignition of the plasma.
3. After the plasma treatment, gently assemble the ridge structure and the top layer onto the bottom layer (*see* **Note 8**).
4. Immerse the assembled device in a beaker with DI water.
5. In order to remove any air bubbles trapped inside the sealed microgrooves area, degas the device immersed beaker in a vacuum chamber for 5–10 min.
6. For sterilization, autoclave the device in water without drying procedure.
7. Take out the device inside a biosafety cabinet and place inside a sterilized petri dish for easy handling.
8. Remove DI water residue in compartments and apply PDL solution to the soma compartment (800 μl) and the axon compartments (200 μl) (*see* **Note 9**).
9. Coat PDL inside a humidified 37 °C incubator overnight.
10. Aspirate PDL solution from all compartments and thoroughly rinse the compartments with sterile DI water 3–5 times with 5–10 min interval (*see* **Note 10**).
11. Apply culture medium to the soma compartment (800 μl) and the axon compartments (200 μl) and store inside a humidified 37 °C incubator overnight before use (*see* **Note 11**). The device is now ready for cell seeding.

3.3 Microsystem Neuron Culture

1. Prepare neurons from E16 rat [15].
2. Aspirate culture medium from the soma compartment.
3. Dilute neurons in 200 μl of NBB27 culture medium with glutamate (cell concentration to match 1,000 cells/mm^2 to the cell culture substrate area) and apply to the soma compartment (*see* **Note 12**).
4. Incubate the neuron seeded PDMS device inside a humidified 5 % CO_2 incubator for 30 min to allow neurons to settle down and attach to the bottom surface.
5. Aspirate culture medium from the six axon compartments.
6. Apply additional 600 μl of culture medium to the soma compartment and add 300 μl of culture medium to each of the six axon compartment.
7. Replace half of culture medium every 3–4 days (*see* **Note 13**).
8. NBB27 with glutamate should be changed to NBB27 without glutamate at DIV 4.

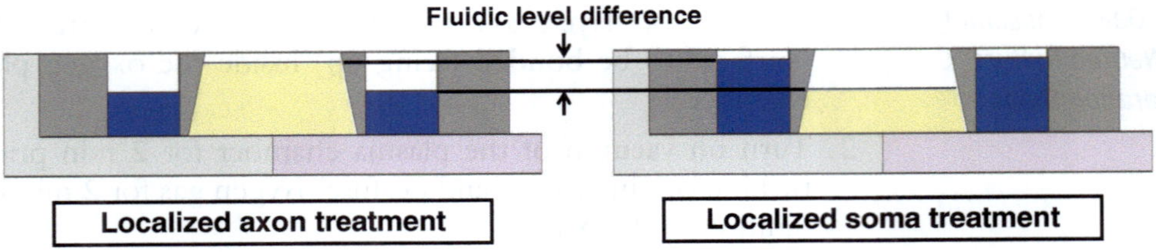

Fig. 4 Illustrations showing fluidic level difference in the case of localized axon treatment (*left*) and localized soma treatment (*right*)

3.4 Localized Biomolecular Treatment

Localized biomolecular treatment either only to the neuronal somata or the isolated axons can be achieved by generating fluidic level difference between the soma compartment and the satellite axon compartments (Fig. 4). Minute culture medium flow driven by the hydrostatic pressure from the compartment having higher fluidic level counters diffusion of the treated biomolecules and confines them within the applied compartment.

3.4.1 Localized Axon Treatment

1. Replace culture medium in the axon compartments with new culture medium (150 μl) containing biomolecule to be treated (*see* **Note 13**).
2. Keep culture medium in the soma compartments to be 800 μl for maintaining fluidic level higher than that of the axon compartments.

3.4.2 Localized Soma Treatment

1. Replace culture medium in the soma compartment with new culture medium (500 μl) containing biomolecule to be treated (*see* **Note 13**).
2. Keep culture medium in the axon compartments to be 400 μl for maintaining fluidic level higher than that of the soma compartment.

3.5 Axon Length Analysis

Axons grown inside the culture platform cannot be seen under light microscopes (i.e., bright field, phase contrast) since axons tend to grow along the corners of the axon-guiding microgrooves. Therefore, cells need to be fluorescently labeled using dye such as Calcein-AM.

3.5.1 Axon Staining with Calcein-AM

1. Aspirate culture medium from the axon compartment (*see* **Note 14**).
2. Add 200 μl of Calcein-AM (1 μM) diluted in culture medium to axon compartments.
3. Incubate cells inside a humidified 5 % CO_2 incubator for 15 min.

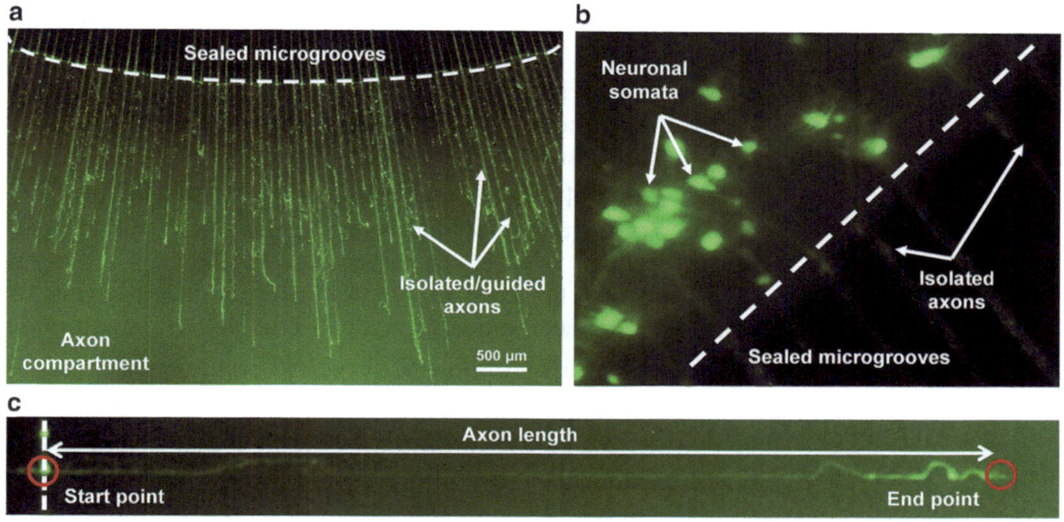

Fig. 5 (**a**) Linear growth of the isolated/guided axons inside the axon compartment at DIV 11. (**b**) Close-up view of the retrograde stained neuronal somata inside the soma compartment having axons isolated inside the axon compartment. (**c**) Measurement scheme of the isolated/guided axon. *White dotted lines* indicate compartment boundaries

4. Aspirate the culture medium in both the soma and the axon compartments and rinse three times with fresh medium.
5. Add 800 and 300 μl of medium to the soma compartment and the axon compartment, respectively.

3.5.2 Axon Length Measurement

When isolated axons are stained with Calcein-AM at DIV 11, linear growth of axons due to the axon-guiding microgrooves can be clearly seen (Fig. 5a). Also, it can be seen by the retrograde staining of the dye that most of the neuronal somata having axons inside the axon compartments are in the vicinity of the sealed microgrooves area (Fig. 5b). Length of axons can be analyzed by measuring the distance from the sealed microgroove outlets and the end of the axons using any image processing software such as ImageJ (Fig. 5c).

4 Notes

1. Piranha cleaning is used to remove organic residues from substrates. Acid piranha is a self-starting reaction which is extremely exothermic. Therefore, the task should always be performed inside a fume hood.
2. When uncross-linked SU-8™ is not properly removed during the development stage, white residue can be observed while rinsing with IPA. In such cases, dry the wafer using pressur-

ized nitrogen gas and further develop in SU-8™ developer (PGMEA).

3. (tridecafluoro-1,1,2,2-tetrahydrooctyl)-Trichlorosilane coating allows the replicated PDMS layer to be easily peeled off from the master mold.
4. The side wall that separates the soma compartment from the axon compartments should be designed to have thicker profile on top (Fig. 1b-inset) to prevent the PDMS wall from being torn during the final top layer replication process (Fig. 3). Alignment marks should be added to the design as appropriate.
5. Improper mixing of polymer base and curing agent can cause unevenly cured PDMS and can contaminate the master mold.
6. Polymerized PDMS surfaces can be easily contaminated by dusts or particles. Wrapping the PDMS devices with parafilm after it has been sufficiently cooled down can prevent contaminations.
7. Intense bubble generation can occur when using a PDMS structure as the replication master mold. Therefore, additional degassing of PDMS mixture prior to pouring it to the PDMS master mold is recommended.
8. Gently assemble the top layer to the bottom layer since sealed microgrooves area can easily be collapsed.
9. In order to properly coat the sealed microgrooves with PDL, generate fluidic level difference between the soma compartment and the axon compartments.
10. PDL inside the sealed microgrooves can be thoroughly rinsed by generating fluidic level difference between the soma compartment and the axon compartments during the rinse process.
11. Incubation of the device allows any potential debris or toxic residue to be removed.
12. Cell numbers can be adjusted according to the experimental design (500–2,000 cells/mm^2 is recommended).
13. Be careful not to damage the cells.
14. Aspirate the culture medium from the outer end of the axon compartments to prevent axons from being damaged.

Acknowledgments

This work was supported by the National Institutes of Health/National Institute of Mental Health grant #1R21MH085267 and by the National Institutes of Health/National Institute of Neurological Disorders and Stroke grant #NS060017.

References

1. Qiu J, Cai D, Filbin MT (2000) Glial inhibition of nerve regeneration in the mature mammalian CNS. Glia 29:166–174
2. Fitch M, Silver J (2008) CNS injury, glial scars, and inflammation: inhibitory extracellular matrices and regeneration failure. Exp Neurol 209:294–301
3. Park KK et al (2008) Promoting axon regeneration in the adult CNS by modulation of the PTEN/mTOR pathway. Science 322: 963–966
4. Campenot RB (1977) Local control of neurite development by nerve growth factor. PNAS 74:4516–4519
5. Hartikka J, Hefti F (1988) Development of septal cholinergic neurons in culture: plating density and glial cells modulate effects of NGF on survival, fiber growth, and expression of transmitter-specific enzymes. J Neurosci 8: 2967–2985
6. Brewer G, Torricelli J, Evege E, Price P (1993) Optimized survival of hippocampal neurons in B27-supplemented neurobasal™, a new serum-free medium combination. J Neurosci Res 35:567–576
7. Goldberg JL et al (2002) Retinal ganglion cells do not extend axons by default: promotion by neurotrophic signaling and electrical activity. Neuron 33:689–702
8. Goldberg JL, Klassen MP, Hua Y, Barres BA (2002) Amacrine-signaled loss of intrinsic axon growth ability by retinal ganglion cells. Science 296:1860–1864
9. Ito D et al (2010) Minimum neuron density for synchronized bursts in a rat cortical culture on multi-electrode arrays. Neuroscience 171: 50–61
10. Winzeler AM et al (2011) The lipid sulfatide is a novel myelin-associated inhibitor of CNS axon outgrowth. J Neurosci 31:6481–6492
11. Taylor AM et al (2005) A microfluidic culture platforms for CNS axonal injury. regeneration and transport. Nat Methods 2: 599–605
12. Park J, Koito H, Li J, Han A (2009) Microfluidic compartmentalized co-culture platform for CNS axon myelination research. Biomed Microdevices 11:1145–1153
13. Park J, Koito H, Li J, Han A (2012) Multi-compartment neuron-glia co-culture platform for localized CNS axon-glia interaction study. Lab Chip 12:3296–3304
14. Park J, Kim S, Park SI, Choe Y, Li J, Han A (2014) A microchip for quantitative analysis of CNS axon growth under localized biomolecular treatments. J Neurosci Methods 221: 166–174
15. Koito H, Li J (2009) Preparation of rat brain aggregate cultures for neuron and glia development studies. J Vis Exp 31:1304

Chapter 8

Organotypic Slice Co-culture Systems to Study Axon Regeneration in the Dopaminergic System Ex Vivo

Claudia Heine and Heike Franke

Abstract

Organotypic slice co-cultures are suitable tools to study axonal regeneration and development (growth or regrowth) of different projection systems of the CNS under ex vivo conditions.

In this chapter, we describe in detail the reconstruction of the mesocortical and nigrostriatal dopaminergic projection system culturing tissue slices from the ventral tegmental area/substantia nigra (VTA/SN) with the prefrontal cortex (PFC) or the striatum (STR). The protocol includes the detailed slice preparation and incubation. Moreover, different application possibilities of the ex vivo model are mentioned; as an example, the substance treatment procedure and biocytin tracing are described to reveal the effect of applied substances on fiber outgrowth.

Key words Organotypic slice co-cultures, Striatum, Prefrontal cortex, Ventral tegmental area, Dopaminergic system

1 Introduction

The target-oriented growth of neuronal processes and fiber projections is essential for both normal physiological conditions and regenerative situations, e.g., after different kinds of injury or neurological diseases. These disorders are, for example, associated with axon growth failures or changes in brain connectivity that alter underlying neuronal circuitries [1, 2]. Therefore, the generation of appropriate models to better understand the development of neuronal fiber pathways and the pathophysiology of different disease patterns are of great importance.

Organotypic slice cultures, which retain the cytoarchitecture of the tissue of origin, have been widely used to examine questions ranging from morphology and development to pathophysiology of brain diseases [3, 4]. They could be derived from different brain regions like the hippocampus [5], mesencephalon [6], cerebellum [7], spinal cord [7, 8], and forebrain [9, 10].

Andrew J. Murray (ed.), *Axon Growth and Regeneration: Methods and Protocols*, Methods in Molecular Biology, vol. 1162, DOI 10.1007/978-1-4939-0777-9_8, © Springer Science+Business Media New York 2014

It is well established that neurons depend on their targets for trophic support [11, 12], so slice co-cultures are essential in the examination of issues such as axonal development and regeneration of fiber projections. Compared with standard single slice cultures, the co-cultures consist of at least two slices of different parts of the same projection system and allow for the assessment of interneuronal responses across brain regions over long distances [4]. Co-culture models may be an interesting alternative for investigations of in vivo situations and have been well established for the entorhino-hippocampal [13–16], cortico-mesencephalic [12, 17–20], cortico-striatal [20–22], nigro-striatal [18, 23], and cortico-spinal [24, 25] preparations.

In this chapter, we describe a protocol for the preparation of organotypic slice co-cultures of the mesocortical and nigrostriatal dopaminergic system. Based on anatomical and functional studies, the dopaminergic system can be divided into three projection systems: the (1) nigrostriatal, (2) mesolimbic, and (3) mesocortical pathway [26]. The neurons that arise from the substantia nigra (SN, A9 cell group) project predominantly to the striatum (STR). Dopaminergic cell bodies originating in the ventral tegmental area (VTA, A10 cell group) innervate several structures like the ventral STR or limbic and cortical areas, e.g., the prefrontal cortex (PFC) [26–28].

In our protocol, slices of the VTA/SN are co-cultured with slices of the PFC or STR using a modification of the membrane interface method of Stoppini and coworkers [29]. The principle is to place slices at the air–medium interface on semipermeable membranes, so they obtain oxygen from above and medium from below. After 10–28 days in vitro (DIV), the VTA/SN + PFC and the VTA/SN + STR cultures retain most of their in vivo characteristics, in that they maintained the shape and localization of, e.g., tyrosine hydroxylase-positive neurons and developed the characteristic fiber network into the prefrontal cortical and striatal subareas [18]. Additionally, as an example of the application possibilities of the model, the substance treatment procedure and biocytin tracing (anterograde tracer, to characterize axonal growth) are described, revealing the effect of applied compounds on neurite outgrowth. After linkage of the co-culturing protocol to a computerized quantification pipeline, a reproducible analysis of the obtained microscopic images is possible regarding the characterization and quantification of the axonal fibers connecting the dopaminergic VTA/SN with the target region [23, 30].

Moreover, using the described co-culture systems, further techniques could be performed: (a) *Immunohistochemistry*, to collect information about the type of developing fibers and the cytoarchitectural organization of the VTA/SN and their target-specific innervations [18, 31]; (b) *Electrophysiology*, to prove the synaptic connectivity and functionality of the projections [32]; or (c) *Gene expression studies*, to reveal the expression on mRNA level (RT-PCR) [31].

2 Materials

2.1 Chemicals and Solutions

Prepare all solutions using ultrapure distilled water and analytical grade reagents. Diligently follow all regulations when working under sterile conditions.

Determine and evaluate the necessary final concentrations of the applied substances. Sterilize the substance solutions using a low binding filter and store the aliquots depending on their respective properties. We recommend subjecting some co-cultures to the vehicle/solvent, for example, artificial cerebrospinal fluid (ACSF) as well as with a positive control substance. Choose the appropriate solvent for the respective substances. Reduce the potential toxicity of some solvents for the slice co-cultures to a minimum.

1. *Standard phosphate buffer (PB)*: 0.1 and 0.2 M, pH 7.3–7.4. The buffer can be obtained from the factory or can be self-made.
2. *ACSF*: Prepare an ACSF solution of the following composition (mM) 126 NaCl, 2.5 KCl, 1.2 NaH_2PO_4, 1.3 $MgCl_2$, and 2.4 $CaCl_2$. Adjust the pH to 7.4 and sterilize with syringe filters (0.2 μm, Sarstedt, Nümbrecht, Germany). Store the solution in the refrigerator in aliquots. The solution is stable for several weeks.
3. *Test substances in question (for example, receptor agonist and antagonist)*: Dissolve the respective substances in the necessary solvent, sterilize with syringe filters (0.2 μm), and store it as stock solution at −20 °C.
4. *Positive control substance (for example, nerve growth factor, NGF)*: Dissolve 100 μg NGF (Sigma-Aldrich) in 10 mL ultrapure water, sterilize with syringe filters (0.2 μm), and freeze at −20 °C as stock solution (10 μg/mL) in aliquots to avoid repeated freezing and thawing.
5. *Preparation medium (p-medium)*: Supplement 100 mL Minimum Essential Medium (MEM, liquid, Invitrogen GmbH) with 1 mL glutamine (200 mM, Invitrogen GmbH, final concentration 2 mM). Adjust the ph to 7.2–7.3 using sterile 1 N sodium hydroxide solution (NaOH, Sigma-Aldrich) or 1 N hydrochloric acid (HCl, Sigma-Aldrich). Before use, add 100 μL of the antibiotic gentamicin (Invitrogen GmbH, final concentration 50 μg/mL). Prepare the medium freshly before every preparation under sterile conditions and cool it during the whole procedure. Caution: NaOH and HCl solutions are corrosive and must be handled with care.
6. *Incubation medium (i-medium)*: Mix 25 mL MEM (twofold concentrated MEM stock solution, prepared from MEM powder medium, Invitrogen GmbH) with 25 mL Hanks' Balanced Salt Solution (HBSS, Invitrogen GmbH) and 25 mL heat-inactivated horse serum (Invitrogen GmbH). Add 23.5 mL

ultrapure water. Supplement the mixture with 1 mL glutamine (200 mM, final concentration 2 mM) and 0.6 mL 7.5 % sodium bicarbonate solution (Sigma-Aldrich, final concentration 0.044 %). Adjust the ph to 7.2 using sterile 1 N NaOH or 1 N HCl. Add 100 μL of the antibiotic gentamicin (final concentration 50 μg/mL) prior usage. Prepare the medium freshly before every preparation under sterile conditions. Without antibiotic, the fresh prepared incubation medium can be stored at 4 °C for 2 weeks. Caution: NaOH and HCl solutions are corrosive and must be handled with care.

7. *Fixation medium (f-medium)*: Dissolve 4 g parformaldehyde (PFA, Merck, Darmstadt, Germany) in 50 mL 0.2 M PB. Moderately heat and stir the solution until it is completely dissolved. Cool off the solution. Afterwards, add 15 mL picric acid solution (Sigma-Aldrich) and 35 mL ultrapure water. The freshly prepared fixation medium can be stored at 4 °C for at least 2 months. Add 0.1 % glutaraldehyde (Serva Electrophoresis GmbH, Heidelberg, Germany) before use. We recommend preparing the solution under a biosafety cabinet. Caution: Paraformaldehyde is harmful, irritant, carcinogenic, and sensitizing. Picric acid is toxic and could be explosive when dry and after shock, friction, fire, or other sources of ignition. Glutaraldehyde is toxic and an irritant, too. All substances must be handled with care. Please use personal protective equipment as required and avoid getting in contact with the skin with the single ingredients and the prepared *f-medium*.
8. *Agarose solution*: Dissolve 1.5 g agarose (Agarose Standard EEO, Serva) in 100 mL distilled water and autoclave the solution. The autoclaved solution can be stored at room temperature for at least 6 months.
9. *3,3'-Diaminobenzidine (DAB)/Ni/Co solution*: Dissolve 5 mg DAB (Sigma-Aldrich) in 7 mL 0.1 M PB. Add 150 μL cobalt chloride (Cobalt (II) Chloride Hexahydrate Solution 1 %, Sigma-Aldrich) and 120 μL ammonium nickel sulfate (Ammonium Nickel (II) Sulfate Hexahydrate Solution 1 %, Sigma-Aldrich). Prepare the solution 5 min before usage and protect it for light. Caution: DAB is harmful and must be handled with care. Please use personal protective equipment (gloves and mask) as required. Dispose waste of DAB solution and DAB containing materials according to regulations.
10. *Hydrogen peroxide (H_2O_2) solution*: Dissolve 10 μL 30 % H_2O_2 (Sigma-Aldrich) in 1 mL 0.1 M PB. Caution: H_2O_2 is corrosive and must be handled with care.

2.2 Animals, Equipment, and Material

1. *Animals*: Neonatal rat pups of postnatal day 1–5 (P1–5) old Wistar rats. The animals must be maintained under standard laboratory conditions (12 h light/dark cycle with food and water ad libitum). All of the animal use procedures must

approved by the Committee of Animal Care and Use of the relevant local governmental body in accordance with the laws of experimental animal protection. The number of animals used should be minimized as well as their suffering.

2. *Equipment and material for tissue culture preparation*: Biosafety cabinet (HERAsafe, Thermo Scientific, Oberhausen, Germany), CO_2 incubator (HERAcell 150, Thermo Scientific), vibratome (Leica, Typ VT 1000S, Nussloch, Germany), binocular stereo-microscope (Olympus SD30, Hamburg, Germany), hotplate, razor blade, tweezers, 6-well cell culture plates (Nalge Nunc International; Thermo Fisher Scientific, Inc., Waltham, MA, USA), membrane inserts (Millicell 0.4 μm, Millipore, Bedford, MA, USA), cell culture dishes (Nunclon™ ⌀ 3.5 cm, Nalge Nunc International), sterile petri dishes (⌀ ca. 10 cm), three beakers (100 mL), automatic micropipettes and sterile tips (Eppendorf, Hamburg, Germany), cyanoacrylate glue (Histoacryl®, B. Braun, Tuttlingen, Germany), 70 % ethanol.
3. *Surgical instruments*: 3–4 extra thin scissors, 1 large scissors, 6 curved tweezers, 1 long straight tweezers, 10 fine spatulas, 2 scalpels (1 with curved and 1 with straight razor blade), glass Pasteur pipette (wide shape), cotton buds soaked in ethanol.
4. *Equipment and material for biocytin tracing and visualization*: fire sealed glass micropipettes, biocytin (Sigma-Aldrich), stiff agarose block, fine brush, Avidin-Biotin-Peroxidase Kit (Vectastain Elite ABC-Peroxidase Kit, Vector Laboratories, Inc., Burlingame, CA, USA), *n*-butylacetate (Merck, Darmstadt, Germany), Entellan® mounting medium (Merck).

3 Methods

Carry out all procedures at room temperature unless otherwise specified. Be careful in working under sterile conditions.

3.1 Slice Preparation and Incubation

3.1.1 Arrangement of the Preparation Reagents

1. Turn on the laminar flow and clean the surface of the biosafety cabinet. Place 6-well cell culture plates, membrane inserts, cell culture dishes, *p-medium* as well as warmed (37 °C) *i-medium* inside the biosafety cabinet.
2. Place 1 mL of the *i-medium* in each well of the 6-well plate and add one membrane insert in each well with the help of tweezers. Subsequently, place the 6-well plate containing the inserts into a CO_2 incubator (standard conditions used for the whole study: 37 °C, 5 % CO_2) for at least 20 min.
3. Place 5 mL of the *p-medium* in every cell culture dish and store it at 4 °C. You need one cell culture dish for every prepared pup.
4. Warm up the agarose solution on a hotplate or in the microwave until the solution is transparent and keep it on a hotplate.

3.1.2 Arrangement of the Preparation Material

1. Clean the surface of the workbench where the vibratome is positioned as well as the buffer tray, the knife holder, and the razor blades of the vibratome very carefully with 70 % ethanol. Every working step performed outside a biosafety cabinet could cause contamination and should be done very diligently and with sterility in mind. Change the razor blade after every session to avoid contamination and to optimize slice quality.
2. Adjust the cutting thickness on the vibratome to 300 μm.
3. Open the box (we recommend to sterilize and to store all surgical instruments in a stainless steel box) containing the surgical instruments under the biosafety cabinet and place 3–4 extra thin scissors, 3 curved tweezers, and 2 spatulas in sterile petri dishes (Ø ca. 10 cm). Proceed in the same way with the following instruments: For the binocular stereomicroscope (one curved tweezers and four fine spatulas); for the vibratome (two curved tweezers, four fine spatulas, one glass Pasteur pipette, two scalpels, with curved and straight razor blade, respectively); and for the decapitation outside of the sterile area (one large scissors, one long straight tweezers). Furthermore place a sterile beaker with 70 % ethanol each, near the petri dishes for the instruments. Maintain sterility during the whole dissection. Clean instruments with ethanol after every use. Do not insert instruments used outside the biosafety cabinet again under the biosafety cabinet without sterilization or cleaning with ethanol.
4. Place ice around the buffer tray of the vibratome as prescribed from the manufacturer to cool the medium while slicing and fill fresh *p-medium* within the tray. Check that all the ethanol has evaporated before filling the buffer tray with the *p-medium*.

3.1.3 Dissection and Slicing of the Brain

1. Pour 8–10 mL of the hot agarose solution in a weighing dish. Work quickly! You have to place the brain in the agarose when the agarose is liquid; be careful that the agarose solution is not too hot otherwise you will harm the brain. It is preferable that the agarose is lukewarm but not solid.
2. Decapitate one pup outside the sterile area with a quick cut at the level of the foramen magnum using a pair of large scissors.
3. Insert long straight tweezers into the eyes to be able to hold the head and to swab the blood from the skin around neck and head with cotton buds soaked in ethanol. Subsequently, move the head under the biosafety cabinet. Maintain sterility during the whole dissection. Use new sterile instruments to cut the skin and the skull. Work carefully and quickly to take out the brain.
4. To remove the skin from the head, cut the skin along the midline of the head with the help of extra thin scissors. Pull the skin to the side with curved tweezers.

5. To remove the brain from the skull, position extra thin scissors close to the skull and cut the cranium bilaterally from the foramen magnum to the front. Perform two additional cuts to the left and to the right side.
6. Peel off the skull with the help of curved tweezers, and remove the brain carefully using a spatula.
7. Place the brain in cold *p-medium* and afterwards in lukewarm agarose solution. Avoid placing the brain in too hot or too stiff agarose as the brain tissue could be damaged.
8. Cool the agarose on ice until the agarose is stiff.
9. To obtain slices of the VTA/SN, perform two coronal cuts using scalpels as shown in Fig. 1Aa (a coronal cut at the level of the pons to remove the cerebellum, a coronal cut caudal of the bulbi olfactorii to remove the remaining cortical part).
10. Glue the agarose embedded tissue block on the specimen plate at the rostral sectional plane, and transfer the specimen plate to the buffer tray.
11. Cut the tissue at the aqueductus mesencephali as well as along the midline with a scalpel (Fig. 1Aa′). Try only to cut the tissue and not the agarose otherwise the agarose block will lose its stability while slicing.
12. Acquire 300 μm thick coronal sections with the help of the vibratome. Use a sterile Pasteur pipette to transfer only the VTA/SN slices (correct regions as indicated by Paxinos and Watson [33]) into sterile cell culture dishes (*see* **Note 1**). Proceed with the protocol in Subheading 3.1.4, **step 1**, subsequently. Up to 8 VTA/SN slices, 4 from each hemisphere, can be obtained from one rat brain. We did not attempt to separate the VTA and the SN.
13. To obtain slices of the forebrain regions (Fig. 1Ab, c), perform the protocol in Subheading 3.1.3, **steps 1–8** in the same way as described for the VTA/SN. Because of the time, we recommend to prepare the cortical and striatal part of the brain from another decapitated pup or to use two vibratomes, simultaneously. It is possible to obtain both regions from the same brain.
14. To obtain slices of the PFC, perform two coronal cuts using scalpels as shown in Fig. 1Ac (a coronal cut to remove the bulbi olfactorii, a coronal cut at forebrain levels to remove the caudal part of the brain.
15. Glue the rostral part of the agarose block on the specimen plate at the fresh sectional plane, and transfer the specimen plate to the buffer tray.
16. Cut the tissue vertical and horizontal to the midline with a scalpel (Fig. 1Ac′). Try only to cut the tissue and not the agarose otherwise the agarose block will lose its stability while slicing.

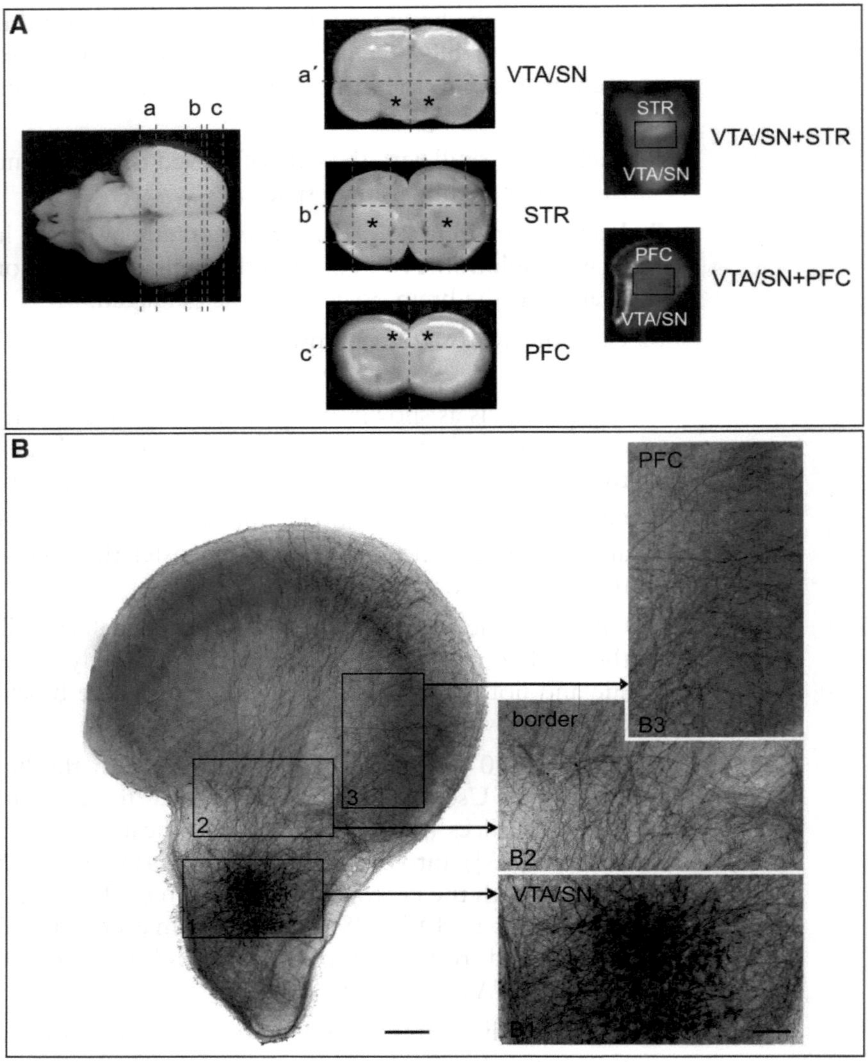

Fig. 1 (**A**) Photographical illustration of the preparation procedure of organotypic slice co-cultures of the mesocortical and the nigrostriatal dopaminergic projection system. Intact brains were removed from 1- to 5-day-old rats. Coronal cuts were made at the level of *a* to isolate the mesencephalon and *b*, *c* the forebrain. Additional cuts were done according to the broken lines to separate the ventral tegmental area/substantia nigra (*a′*, VTA/SN), the striatum (*b′*, STR), and the prefrontal cortex (*c′*, PFC). Selected coronal sections (300 μm thick) were placed on membranes as VTA/SN + STR and VTA/SN + PFC co-cultures. (**B**) Overview of a co-culture VTA/SN + PFC (DIV10) after biocytin tracing of the VTA/SN and visualization with DAB; the images present the black-labelled cell bodies in (*1*) the VTA/SN and their outgrowing fibers, passing (*2*) the border and growing into (*3*) the PFC. (*B1–B3*) Higher magnifications of (*1–3*). (Scale bars: *B* = 500 μm; *B1–B3* = 200 μm)

17. Obtain 300 μm thick coronal sections with the help of the vibratome. Use a sterile Pasteur pipette to transfer the PFC slices (correct regions as indicated by Paxinos and Watson

[33]) into sterile cell culture dishes (*see* **Note 1**). Proceed with the protocol in Subheading 3.1.4, **step 1**. Up to 8 PFC slices, 4 from each hemisphere, could be obtained on average from one rat brain.

18. To obtain slices of the STR, perform two coronal cuts as shown in Fig. 1Ab (a coronal cut at forebrain levels to remove the caudal part of the brain, a coronal cut to separate the PFC region).
19. Using a scalpel cut the tissue as indicated in Fig. 1Ab′. Try only to cut the tissue and not the agarose otherwise the agarose block will lose its stability while slicing.
20. Transfer only the STR slices (correct regions as indicated by Paxinos and Watson [33]) to sterile cell culture dishes (*see* **Note 1**) and proceed with the protocol in Subheading 3.1.4, **step 1**. Up to 8 STR slices, 4 from each hemisphere, can be obtained on average from one rat brain. We recommend changing the ***p-medium*** in the buffer tray at least for every other animal.

3.1.4 Plating of Slices as Co-culture

1. Place the 6-well cell culture plate containing the membrane inserts in the biosafety cabinet. Check quality of obtained slices under the binocular stereomicroscope (*see* **Note 2**). Remove remaining meninges from brain slices with the help of tweezers and spatula. Unremoved meninges decrease the quality of the slices. Do not place more than 3–4 co-cultures on each membrane.
2. Transfer selected VTA/SN slices on the membranes of the inserts with the help of a spatula (*see* **Note 3**). Afterwards, put the 6-well plate back to the CO_2 incubator until the cortical or striatal slices need to be put on the membrane.
3. Proceed in the same way with the PFC or the STR slices (*see* **Note 3**). Place one slice 0.5–1 mm away from each VTA/SN section, to receive VTA/SN + PFC or VTA/SN + STR co-cultures, respectively. Use a fine spatula for the reorientation of the slices if necessary. Return the 6-well plate back to the CO_2 incubator. Store the cultures in the CO_2 incubator.
4. Change the ***i-medium*** directly after finishing the preparation procedure and once more on the next day. Thereafter, change the ***i-medium*** thrice weekly. Prior to medium change, warm the ***i-medium*** (37 °C), adjust the pH to 7.2 if necessary, and add the antibiotic gentamicin (final concentration 50 μg/mL). While changing the media, observe the survival of the cultures with the help of the binocular stereomicroscope. Discard co-cultures that show signs of microbial contamination (*see* **Note 4**).

3.2 Substance Treatment and Biocytin Tracing

3.2.1 Substance Treatment Procedure of the Co-cultures

1. For substance treatment, add the respective substance to 1 mL warm *i-medium*. Place the 6-well plate under the biosafety cabinet, and replace the old medium with fresh medium containing the substances. Observe if precipitation of the substance occurs while applying the *i-medium* in which case final concentrations are uncertain.
2. Execute the treatment at least four times within the incubation period every time the incubation medium is changed (Fig. 2) (*see* **Note 5**).

3.2.2 Tracing Procedure and Fixation of the Co-cultures

1. Perform the tracing procedure 2 days before fixation (for details *see* Fig. 2). Turn on the laminar flow, and clean the surface of the biosafety cabinet with ethanol.
2. Place a small amount of biocytin crystals in a sterile cell culture dish. We recommend using glass dishes instead of plastic dishes due to lower electrostatic attraction of the glass.

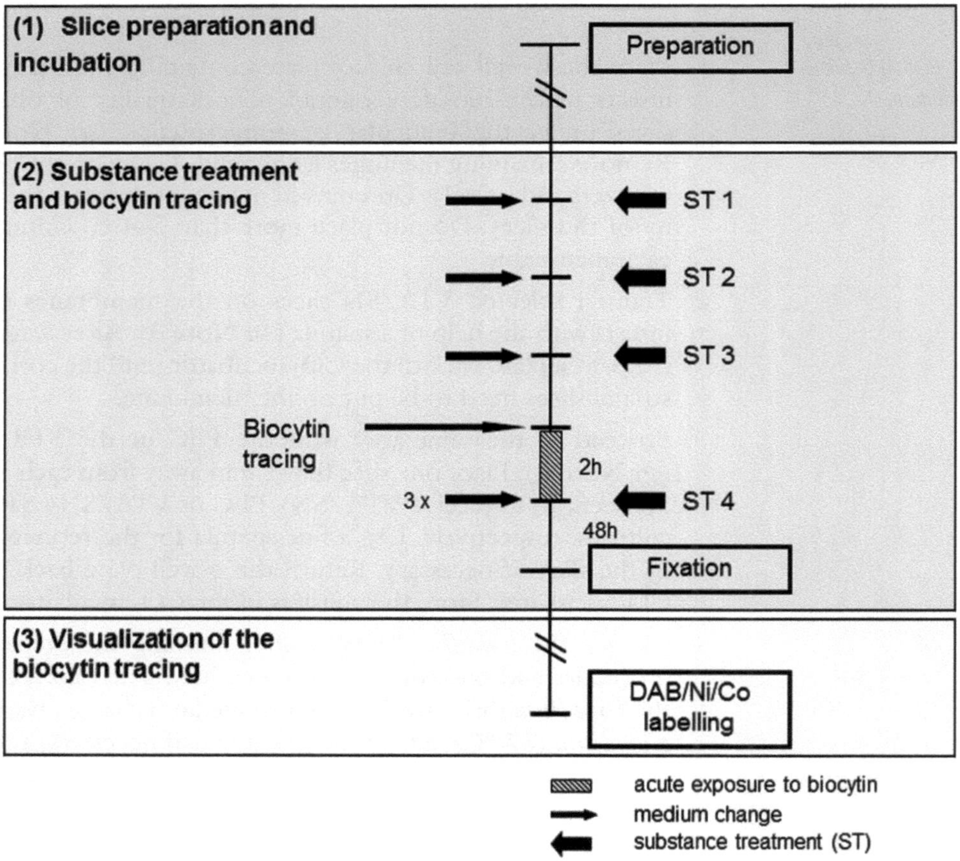

Fig. 2 Chronological scheme to illustrate the experimental procedures for substance treatment and biocytin tracing. During the incubation period of the cultures, the substance treatment (ST) was executed four times (*ST1–ST4*) on the days the medium was changed. Biocytin tracing was performed as described and after 24 h the co-cultures were fixed and could be analyzed

3. Position the 6-well plate in the biosafety cabinet and open the plate.
4. With the aid of a binocular microscope, use a fire sealed glass micropipette to put small equal biocytin crystals onto the VTA/SN part of the co-culture. Place the crystals very carefully onto the tissue (*see* **Note 6**).
5. Put the 6-well plate back to the CO_2 incubator for the next 2 h to allow the uptake of the biocytin crystals.
6. Afterwards, place the 6-well plate in the biosafety cabinet. Lift each insert with sterile tweezers, tilt the inserts, and apply the warm *i-medium* to the top of the inserts. Take off the entire medium from the membranes and exchange the medium in the 6-well plate under the inserts (*see* **Notes 6** and 7). Place the inserts back into the plate and return the 6-well plate to the CO_2 incubator. Repeat the procedure thrice at intervals of 10 min.
7. To allow the anterograde transport of the tracer, reincubate the cultures with warmed *i-medium* containing the substances to be examined. Proceed with the protocol in Subheading 3.2.1, **step 1**.
8. After 48 h remove the media and fix the cultures by adding 2–3 mL *f-medium* inside and below the membrane inserts (*see* **Note 7**). We recommend performing the fixation step under a biosafety cabinet. Repeat the procedure four times at intervals of 30 min.
9. Afterwards, rinse the inserts thrice with 0.1 M PB (*see* **Note** 7) and place the 6-well plate into the refrigerator. You could store the cultures for at least 1 week. We recommend proceeding with the protocol in Subheading 3.3, **step 1** on the next day.

3.3 Visualization of the Biocytin Tracing

We recommend slicing the co-cultures prior to visualization of the biocytin tracing. Otherwise, it is possible to detach the co-cultures from the membrane as "whole mount" using a fine brush.

1. Glue a stiff agarose block on the specimen plate of the vibratome and transfer the specimen plate to the buffer tray containing 0.1 M PB. Cut a plane surface.
2. Cut the membrane of the membrane inserts around the co-culture and glue the co-culture containing membrane piece on the plane surface of the agarose block.
3. Perform 50 μm slices of the co-cultures with the vibratome. Transfer the co-culture slices to a 24-well plate filled with 0.1 M PB using a fine brush.
4. Incubate the co-cultures with an Avidin-Biotin-Peroxidase Kit overnight at 4 °C. Follow the manufacturer's instructions for reagent concentrations and dilution.
5. Afterwards, wash the co-cultures thrice at intervals of 10 min with 0.1 M PB.

6. Incubate the co-cultures with *DAB/Ni/Co solution*.
7. Add 1–2 μL of the *H_2O_2 solution* to start the reaction and incubate the co-cultures for 4–8 min, approximately. Monitor the intensity of the black-blue staining product. Use the same conditions for every study.
8. Stop the reaction by rinsing the co-cultures with 0.1 M PB when staining intensity is achieved. Repeat process of washing thrice at intervals of 5–10 min.
9. Mount the co-cultures on glass slides using a fine brush. Dehydrate dried stained sections in a series of graded ethanol as well as twice in *n*-butylacetate. Cover slip slides with a mounting medium, for example, Entellan®. Caution: Entellan® is flammable and harmful. Keep away from sources of ignition. Wear suitable protective clothing and gloves.

After DAB/Ni/Co labelling, microscopic images can be obtained and analyzed, for example, to characterize the growth promoting effects of substance treatment. Examples of images of the border region are given in Fig. 3.

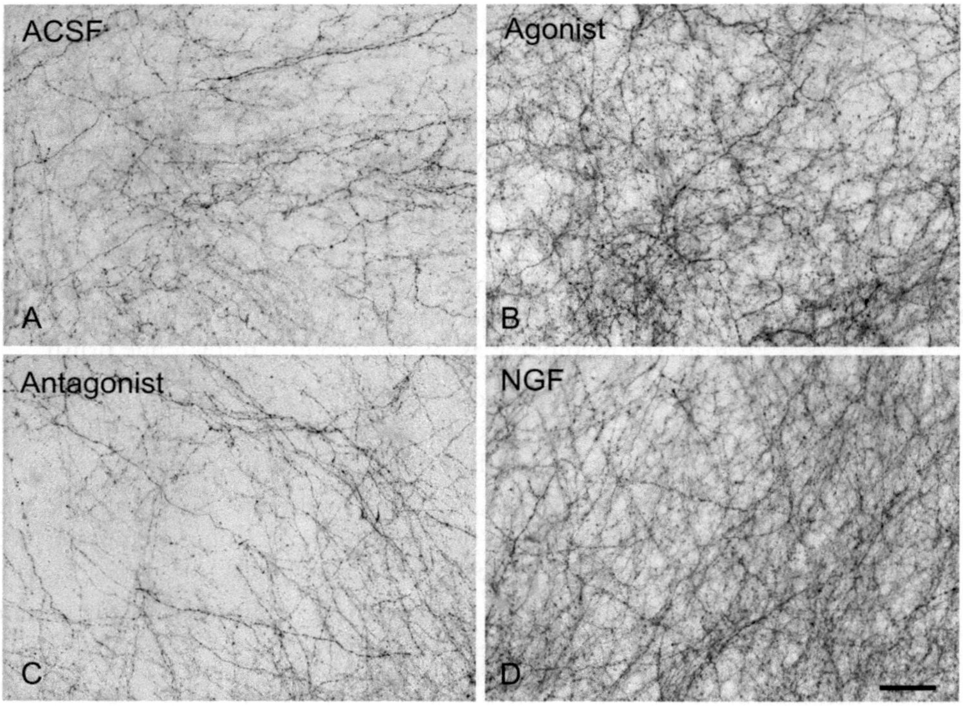

Fig. 3 Examples of co-culture images of the border after substance application (**A**: ACSF, **B**: agonist, **C**: antagonist, **D**: NGF), biocytin tracing and DAB/Ni/Co labelling. (Scale bars: **A**–**D** = 20 μm)

Further application possibilities could be:

1. Morphological or anatomical characterization using *Immunohistochemistry or Immunofluorescence labelling.*
2. Protein expression analysis using *Western blot technique.*
3. Gene expression analysis using *quantitative PCR or Array studies.*
4. Detailed fiber growth analysis using *Tracing techniques and computer-based Image analysis.*
5. Toxicity testing (e.g., after substance treatment) using *Enzyme activity measurements*, for example, release of lactate dehydrogenase into the culture medium.

4 Notes

1. Transfer the slices carefully into the cell culture dishes. Avoid damaging the slices. Pay attention that the brain slices are uniform and do not vary in their thickness. If they vary verify that the razor blades are well adjusted and that they remain free of tissue glue, which can cause inconsistent slicing. If you are not able to solve the problem, we recommend replacing the razor blade with a new and clean one.
2. Discard damaged slices because of their diminished survival. Moreover necrotic slices can influence intact slices. If the slices are folding (becoming wavy, corrugating) in the cell culture plate, check also the pH of the *p-medium* and adjust it to 7.2–7.3 if necessary. Work quickly; a long procedure diminishes the quality of the slices.
3. Work quickly in placing the slices on membranes; a long procedure diminishes the quality of the slices. Work carefully; do not damage the obtained slices or the surface of the membranes.
4. The slices should be transparent after 3–4 days in culture. Otherwise, they are not developing or growing very well because of contamination, damage, or inadequate nutrition. Repeat the preparation with new media and sterilized surgical material and equipment. Maintain sterility during the whole procedure! Do not place more than 3–4 co-cultures on each membrane and ensure that the amount of *i-medium* is 1 mL per well.
5. If the co-cultures become white after some days of substance treatment, it could be that the test substances or the vehicle/ solvent are toxic in the used concentration or that the co-cultures are contaminated. Repeat the preparation with new media and sterilized surgical material and equipment. Sterilize all solutions and reagents. Check the toxicity of the substances and solvents.

6. Place the crystals very precisely on the VTA/SN. Do not bring the untraced PFC or STR regions into contact with the remaining biocytin crystals or biocytin-containing *i-medium*. Work thoroughly and carefully; do not damage the slices or the surface of the membranes. If black-labelled biocytin traced cell bodies and fibers could be found all over the co-culture after visualization with DAB, the application of the tracer was imprecise or the tracer spread while washing the co-cultures after the biocytin incubation. Repeat the preparation. Pay attention to avoid contact to the PFC or the STR parts of the co-culture while placing the tracer and washing after the incubation.
7. Be careful while washing the slices. Apply the media (e.g., *i-medium*, *f-medium*, or PB) very gently from the top and do not place the media directly on the co-cultures, as they might become detached from the membranes. Another reason why the slices may detach from the membrane while washing could be that the slices are contaminated or necrotic. If so, repeat the preparation with new media and sterilized surgical material and equipment. Maintain sterility during the whole procedure!

Acknowledgments

The authors would like to thank K. Krause, K. Becker, K. Sygnecka, and J. Ding for technical assistance. This study was supported by the Deutsche Forschungsgemeinschaft (FR 1253/3-2) and by the German Federal Ministry of Education and Research (BMBF 1315883).

References

1. Chase TN, Oh JD, Blanchet PJ (1998) Neostriatal mechanisms in Parkinson's disease. Neurology 51:30–35
2. Sesack SR, Carr DB (2002) Selective prefrontal cortex inputs to dopamine cells: implications for schizophrenia. Physiol Behav 77:513–517
3. Gähwiler BH, Capogna M, Debanne D et al (1997) Organotypic slice cultures: a technique has come of age. Trends Neurosci 20:471–477
4. Cho S, Wood A, Bowlby MR (2007) Brain slices as models for neurodegenerative disease and screening platforms to identify novel therapeutics. Curr Neuropharmacol 5:19–33
5. Strassburger M, Braun H, Reymann KG (2008) Anti-inflammatory treatment with the p38 mitogen-activated protein kinase inhibitor SB239063 is neuroprotective, decreases the number of activated microglia and facilitates neurogenesis in oxygen-glucose-deprived hippocampal slice cultures. Eur J Pharmacol 592:55–61
6. Østergaard K, Schou JP, Zimmer J (1990) Rat ventral mesencephalon grown as organotypic slice cultures and co-cultured with striatum, hippocampus, and cerebellum. Exp Brain Res 82:547–565
7. Birgbauer E, Rao TS, Webb M (2004) Lysolecithin induces demyelination in vitro in a cerebellar slice culture system. J Neurosci Res 78:157–166
8. Cho J-S, Park H-W, Park S-K et al (2009) Transplantation of mesenchymal stem cells enhances axonal outgrowth and cell survival in an organotypic spinal cord slice culture. Neurosci Lett 454:43–48
9. Vergni D, Castiglione F, Briani M et al (2009) A model of ischemia-induced neuroblast activation in the adult subventricular zone. PLoS ONE 4:e5278. doi:10.1371/journal.pone.0005278
10. Tanvig M, Blaabjerg M, Andersen RK et al (2009) A brain slice culture model for studies of endogenous and exogenous precursor cell

migration in the rostral migratory stream. Brain Res 1295:1–12

11. Plenz D, Kitai ST (1996) Organotypic cortex-striatum-mesencephalon cultures: the nigrostriatal pathway. Neurosci Lett 209:177–180
12. Gomez-Urquijo SM, Hökfelt T, Ubink R et al (1999) Neurocircuitries of the basal ganglia studied in organotypic cultures: focus on tyrosine hydroxylase, nitric oxide synthase and neuropeptide immunocytochemistry. Neuroscience 94:1133–1151
13. Heimrich B, Frotscher M (1993) Slice cultures as a model to study entorhinal-hippocampal interaction. Hippocampus 3:11–17
14. Heine C, Heimrich B, Vogt J et al (2006) P2 receptor-stimulation influences axonal outgrowth in the developing hippocampus in vitro. Neuroscience 138:303–311
15. Del Rio JA, Heimrich B, Borrell V et al (1997) A role for Cajal-retzius cells and reelin in the development of hippocampal connections. Nature 385:70–74
16. Del Turco D, Deller T (2007) Organotypic entorhino-hippocampal slice cultures-a tool to study the molecular and cellular regulation of axonal regeneration and collateral sprouting in vitro. Methods Mol Biol 399:55–66
17. Jaumotte JD, Zigmond MJ (2005) Dopaminergic innervation of forebrain by ventral mesencephalon in organotypic slice co-cultures: effects of GDNF. Brain Res Mol Brain Res 134:139–146
18. Franke H, Schelhorn N, Illes P (2003) Dopaminergic neurons develop axonal projections to their target areas in organotypic co-cultures of the ventral mesencephalon and the striatum/prefrontal cortex. Neurochem Int 42:431–439
19. Kress GJ, Reynolds IJ (2005) Dopaminergic neurotoxins require excitotoxic stimulation in organotypic cultures. Neurobiol Dis 20: 639–645
20. Kristensen BW, Noraberg J, Jakobsen B et al (1999) Excitotoxic effects of non-NMDA receptor agonists in organotypic corticostriatal slice cultures. Brain Res 841:143–159
21. Plenz D, Aertsen A (1996) Neural dynamics in cortex-striatum co-cultures-II. Spatiotemporal characteristics of neuronal activity. Neuroscience 70:893–924
22. Plenz D, Aertsen A (1996) Neural dynamics in cortex-striatum co-cultures-I. Anatomy and electrophysiology of neuronal cell types. Neuroscience 70:861–891
23. Heine C, Sygnecka K, Scherf N et al (2013) Phosphodiesterase 2 inhibitors promote axonal outgrowth in organotypic slice co-cultures. Neurosignals 21:197–212
24. Oishi Y, Baratta J, Robertson RT et al (2004) Assessment of factors regulating axon growth between the cortex and spinal cord in organotypic co-cultures: effects of age and neurotrophic factors. J Neurotrauma 21:339–356
25. Nakamae T, Tanaka N, Nakanishi K et al (2009) Chondroitinase ABC promotes corticospinal axon growth in organotypic cocultures. Spinal Cord 47:161–165
26. Björklund A, Dunnett SB (2007) Dopamine neuron systems in the brain: an update. Trends Neurosci 30:194–202
27. Fallon JH, Loughlin SE (1995) Substantia nigra. In: Paxinos G (ed) The rat nervous system, 2nd edn. Academic, San Diego, CA, pp 215–237
28. Lindvall O, Björklund A (2004) Cell therapy in Parkinson's disease. NeuroRx 1:382–393
29. Stoppini L, Buchs PA, Muller D (1991) A simple method for organotypic cultures of nervous tissue. J Neurosci Methods 37:173–182
30. Scherf N, Kuska JP, Heine C et al (2008) Segmentation of axonal fibres in tissue slices. In: Tolxdorff T et al (eds) Bildverarbeitung für die Medizin 2008: algorithmen-systeme-anwendungen. Springer, Berlin, pp 178–182
31. Heine C, Wegner A, Grosche J et al (2007) P2 receptor expression in the dopaminergic system of the rat brain during development. Neuroscience 149:165–181
32. Dossi E, Heine C, Servettini I et al (2012) Functional regeneration of the ex-vivo reconstructed mesocorticolimbic dopaminergic system. Cereb Cortex 23:2905–2922. doi:10.1093/cercor/bhs275
33. Paxinos G, Watson C (1998) The rat brain in stereotaxic coordinates, 4th edn. Academic, San Diego, CA

Chapter 9

Monitoring Neuron and Astrocyte Interactions with a 3D Cell Culture System

James B. Phillips

Abstract

Methods are described for the generation and analysis of 3D co-culture models in which astrocyte and neuronal behavior can be studied. Cells may be obtained from a variety of sources, then cultured within collagen hydrogels to explore cellular responses and interactions in response to substances under test or under conditions that mimic physiological or pathological environments. Cell populations are labelled then either mixed within gels or arranged as separate adjacent populations, with further options including directing the self-alignment of cells to form anisotropic 3D cultures. Immunofluorescence staining and confocal microscopy can be used to capture image data from 3D structures and detailed protocols are provided for obtaining reliable results. Finally, 3D image analysis of confocal microscopy data is discussed, providing guidance on how astrocyte and neuronal features can be quantified.

Key words Neuron, Astrocyte, Neurite, 3D cell culture, Confocal microscopy, Collagen hydrogel, Neural tissue engineering, Reactive gliosis

1 Introduction

Neurons and astrocytes are intimately related in the CNS and their interactions during physiology and pathology are the focus of much research [1, 2]. Three-dimensional (3D) cell culture models provide useful tools for neuroscience because they can recreate aspects of the spatial and mechanical environment of the CNS, allowing cells to interact and respond naturally in a highly controlled and easily monitored setting [3]. This approach, whereby artificial model tissues are engineered from cells and extracellular matrix in vitro, is particularly beneficial where neural cells are concerned since it overcomes some of the limitations associated with traditional monolayer culture [4, 5]. For example neurons in culture respond differently to 2D and 3D environments [6, 7], and astrocytes adopt a less reactive phenotype in 3D gels compared to stiff monolayer substrates [8, 9]. Furthermore, the glial scar and

Andrew J. Murray (ed.), *Axon Growth and Regeneration: Methods and Protocols*, Methods in Molecular Biology, vol. 1162,
DOI 10.1007/978-1-4939-0777-9_9, © Springer Science+Business Media New York 2014

other interfaces associated with inhibition of neuronal growth within the CNS (e.g., the dorsal root entry zones in spinal cord tissue) are 3D structures, so to recreate their functional architecture in vitro would logically require a 3D approach.

In order for 3D co-cultures of neurons and astrocytes to become widely adopted as platforms for neuroscience research, it is important to establish robust protocols both for the preparation and maintenance of cultures, the application of treatments (for example, to simulate damage or disease), and the monitoring of responses. A wide variety of cell sources and extracellular matrices are available for generating co-cultures of neurons and glia, ranging from reaggregation of dissociated brain tissue or differentiated embryonic stem cells in drops of media [10, 11] or as spheroids in continuously agitated suspension [12], through to the trapping of cells or explanted tissue within gels [8, 13, 14], and the organization of cellular gels into tissues with more complex 3D architecture to recreate interfaces, anisotropy or to simulate mechanical features [15–18]. Here we describe approaches for seeding astrocytes and neurons within collagen gels and organizing the subsequent co-cultures in order to study the reciprocal effects of mixed or adjacent populations of neurons and astrocytes, including the use of anisotropic cultures to align neural cell growth. Protocols describe the establishment and maintenance of these cultures, then focus on monitoring cellular behavior within the models. The protocols for monitoring neuronal length and orientation and astrocyte reactivity are intended to be applicable to a range of 3D neural culture systems using various sources of cells and matrices, providing researchers with practical guidance for applying quantitative analysis to their system of choice. There are numerous sources of astrocytes and neurons available for use in culture systems and the selection of these will depend on the specific research question (e.g., appropriate species, purity, availability, developmental stage, genetic alteration). It is important to be able to distinguish between the astrocytes and the neurons, and this can be achieved by immunolabelling cell-specific proteins or pre-labelling the cell populations with a fluorescent marker (e.g., GFP or CellTracker™). It is also important to ensure that the culture medium is suitable for supporting growth of both neurons and glia in co-culture and that the cell seeding density is optimized for both populations. The methods described here are based on experiences using primary neonatal rat astrocytes and adult rat DRG neurons, with the astrocytes cultured from animals that are genetically altered to express GFP (β-actin reporter), in co-culture experiments to explore neuron–astrocyte interactions over periods of up to 15 days in collagen gels [8, 15, 17].

2 Materials

2.1 3D Cell Cultures

1. Type 1 collagen (2 mg/ml in 0.6 % acetic acid) and neutralizing solution (e.g., RAFT reagent kit, TAP Biosystems).
2. 10× Minimum essential medium (MEM).
3. DMEM with 10 % fetal calf serum and penicillin/streptomycin (100 U/ml).
4. Culture moulds. 3D gels can be cast within standard polystyrene cell culture plates and dishes, imaging chambers, or bespoke interface/alignment moulds (Fig. 1).

2.2 Immunostaining

1. Paraformaldehyde (4 %) is stored and used at 4 °C for fixation.
2. Triton X-100 is used at 0.5 % in PBS for permeabilization.
3. Various antibodies are used for detection of neurons and astrocytes and a range of commercial sources are available. For general assessment of neurites in culture immunolabelling of βIII-tubulin works well (e.g., monoclonal anti-β-tubulin III, 1:400, Sigma T8660), whereas for astrocytes glial fibrillary acidic protein (GFAP) is used widely (e.g., polyclonal rabbit anti-GFAP, 1:300, DAKO Z0334). Secondary antibodies should be chosen carefully to allow multiple signals to be resolved using the available equipment, with minimal cross-talk and background fluorescence (e.g., the Dylight® range from Vector Labs).

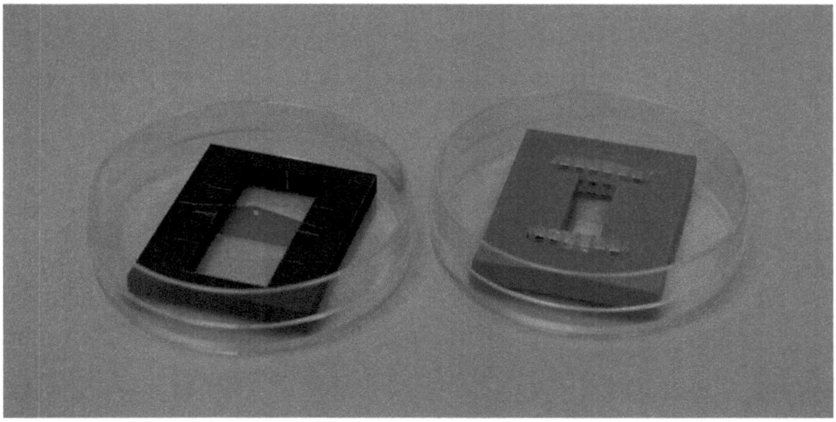

Fig. 1 Examples of an interface (*left*) and an alignment (*right*) mould for making 3D co-cultures from cellular hydrogels. The examples shown are relatively large moulds designed to fit in 60 mm diameter petri dishes. The interface mould has a stainless steel divider that separates sections of the chamber during casting, and can be removed to enable integration during setting [17]. The alignment mould contains nylon mesh structures at each end to provide tethering points to facilitate cellular self-alignment [19]

2.3 Microscopy and Analysis

1. Standard glass slides and cover slips can be combined with various imaging chambers (e.g., CoverWell™) for microscopy of 3D cultures after removal from moulds, or cultures can be imaged in situ if cast within plates or moulds with appropriate transparency.
2. Various microscopy options are available for capturing images from fluorescently labelled 3D cultures. Confocal microscopy can be performed using inverted or upright configurations. A range of different objective lens options is useful; water-immersion lenses in particular offer long working distances that can facilitate monitoring within deep 3D constructs.
3. Image analysis software can be used to quantify cellular features and various options are available. While many microscope systems include software for processing and analyzing 3D image datasets, it can be advantageous to use a separate computer system for analysis. Commercially available 3D imaging software (e.g., Volocity®) can be used to visualize and quantify images from a range of different microscopy platforms.

3 Methods

The specific cell types and the size, shape, seeding density, and complexity of the 3D culture system will be determined by the research question and availability of resources (both for construction of cultures and subsequent analysis). The simplest 3D co-cultures involve mixing the two cell populations within one gel or seeding one cell type onto the surface of a 3D gel containing the other [15, 20]. More advanced systems involve seeding separate cell populations in adjacent 3D compartments or organizing cells in response to mechanical cues [17]. Organizing cells so that they are aligned can make subsequent analysis simpler and faster; measuring random neuronal extension in three dimensions can be considerably more difficult than measuring straight parallel neurites.

3.1 Cell-Seeded Collagen Gels

1. Mix 80 % collagen solution with 10 % MEM (v/v) then neutralize by adding the appropriate volume of neutralizing reagent (*see* **Note 1**). Keep all reagents cold (4 °C) throughout and swirl carefully to ensure thorough mixing without introducing bubbles.
2. Add 10 % v/v cell suspension to provide appropriate cell seeding density (*see* **Note 2**) and mix thoroughly.
3. Quickly transfer the solution to the culture plates or moulds then incubate at 37 °C for 10 min to set.
4. If an additional cell population is to be included as a simple extra layer, then it can be seeded onto the surface of the gel at this stage [20].

5. Add culture medium gently to the top of the gel without causing disruption to the structure. Maintain in culture, changing medium regularly. Treatments such as soluble agents can be administered via the media; ensure that the volume of gel is taken into account when calculating concentrations.
6. To make an interface culture with separate adjacent populations of astrocytes and neurons, make two separate gels as described above and transfer to adjacent wells of an interface culture mould (e.g., Fig. 1) at 4 °C. After 2 min, carefully remove the divider and allow the gels to integrate and set at 21 °C for 10 min then 37 °C for 20 min. Controlling temperature in this way allows the gels to integrate before setting, so the astrocytes and neurons are able to interact with no physical barrier [17].
7. To make an anisotropic 3D co-culture, the astrocyte component can be used to self-align within a tethered rectangular gel. Neurons can either be added to the surface of the aligned construct or mixed with the astrocytes prior to gelling, and the resulting growth of neurites will be guided by the aligned astrocyte environment [15]. To achieve cellular self-alignment, set the gels within rectangular moulds that have attachment points for tethering the collagen at opposite ends, as in Fig. 1 [19] (*see* **Note 3**).
8. Cellular collagen gels can be maintained in culture at 37 °C (5 % CO_2) in a humidified incubator for sufficient time for neurite growth and/or changes in astrocyte reactivity to develop. Cultures must be monitored frequently to check for signs of gel contraction, and media can be replenished as required. Undesirable gel contraction as a result of cell–matrix interaction can be avoided by decreasing the astrocyte seeding density or increasing the collagen density (by increasing the starting concentration of protein or by using plastic compression to remove some interstitial fluid, *see* **Note 4**).
9. If required, continuous monitoring of cultures can be performed by sampling culture media (e.g., to detect levels of growth factors/cytokines via subsequent ELISA) or through the use of video microscopy to generate time-lapse image sequences [8]. Collagen gels are transparent which permits visualization of live cells under phase contrast optics, and the presence of a fluorescent marker can be used to identify cell populations and conduct time-lapse fluorescence microscopy.

3.2 Immunostaining

A range of microscopic and molecular analyses can be used to detect cellular features and interactions at the end of each experimental incubation. For example cells can be lysed in situ for mRNA extraction or Western blotting, or cells can be collected from gels following collagenase incubation [19]. Cellular collagen gels can also be embedded and sectioned in the same way as tissues for histological or ultrastructural analysis [21]. This protocol describes the use of immunofluorescence staining and confocal microscopy of intact cellular gels to monitor the spatial interactions between neurons and astrocytes in 3D culture models.

1. Remove medium and wash briefly with PBS prior to fixation of gels in 4 % paraformaldehyde at 4 °C for at least 3 h (ideally gels should be left overnight) (*see* **Note 5**).
2. For immunostaining of intracellular components, an initial permeabilization step can be carried out using Triton X-100 for 30 min, followed by blocking with 5 % species-specific serum for 30 min.
3. Standard immunostaining protocols developed for monolayer cultures can be applied, but incubation times can be doubled and washes between incubations should be thorough (e.g., 3×5 min washes in 20 ml PBS). A marker such as Hoechst 33258 or DAPI can be useful for identifying cell nuclei.
4. Immunostained gels can be stored in PBS at 4 °C (protected from light) prior to microscopy.

3.3 Microscopy and Analysis

Fluorescently labelled cells in collagen gels can be visualized using a variety of different imaging modalities. Confocal microscopy provides a convenient method for sampling 3D regions of gels and, when conducted with standardized protocols, can provide data for quantification of cell responses. This section describes approaches that can be used to gather and quantify data relating to astrocyte morphology and neurite growth, but the concepts and techniques are broadly applicable to a wider range of 3D image analysis requirements. The development of a robust sampling protocol can benefit from pilot experiments to explore the optimal microscope configuration and sample positioning. Gels can be imaged in their plates or moulds or transferred to imaging chambers or slides, cut into pieces or maintained intact. It is important to note that neural cells can be sensitive to small differences in the physical properties of the extracellular matrix, and it is inevitable that localized areas of altered stiffness or tension will develop and influence cell behavior [22]. For example, astrocytes at the edges of gels tend to exhibit a more reactive phenotype than those elsewhere, and neuronal growth in vitro can be influenced by gradients of stiffness [23]. Sampling protocols must take account of patterns of cell behavior that may be a consequence of the physical environment either by avoiding or selecting particular zones [15].

1. Mount the gel in an appropriate manner—this will depend on the dimensions of the gel and the microscope configuration. For inverted microscopes, gels can be analyzed through the bottom of the culture well/mould or placed on a thin glass slide. Imaging chambers can be used to mount gels for upright microscopes with oil immersion lenses, whereas water

immersion lenses provide the option of keeping the sample in solution into which the lens can dip. In all cases ensure the gel cannot dry out or move around during microscopy, and that the 3D gel structure is not distorted (e.g., avoid squashing it under a cover slip).

2. Choose a suitable objective lens based on the minimum resolution required to resolve the cellular detail required for the research. Typically 20× or 40× objectives are used to assess neuronal and astrocyte features. Higher power lenses reduce the size of the field so more sampling regions are required to monitor the same number of cells, whereas lower power lenses are not sufficient to resolve fine neurite or astrocyte processes.
3. Define the depth of the sampling region (*z*-distance), the number of sampling regions per gel and their positions (*see* **Note 6**).
4. Standardize the laser power and scanning parameters to ensure consistency is maintained. It is useful to conduct pilot experiments with the experimental group that provides the weakest signal (e.g., the least reactive astrocytes or the thinnest neurites) to ensure the features at the lower limit are detectable.
5. Set the gain and offset to give an optimal signal that allows the full range of required fluorescence intensities to be detected. Ensure that the highest intensity features in the experimental group with the strongest signal do not become saturated (*see* **Note 7**).
6. Randomize the gels to be sampled and consider coding them to avoid bias. Apply an identical sampling protocol to each gel and take care to image all the gels at a similar time after staining to avoid changes in signal influencing the results.
7. Astrocytes in 3D collagen gels undergo striking alterations in size, shape, and expression of proteins such as GFAP as well as proliferating in response to stimuli, so there are various options available for assessing their behavior based on the confocal microscope data. A useful and straightforward approach is to use a 3D image analysis program to measure the volume of fluorescence associated with GFP, GFAP staining, etc. (Fig. 2a), then determine the number of cells to which this volume relates, in order to express the fluorescence signal as a volume per cell (*see* **Note 8**). More advanced image analysis programs can quantify the shape of cells (e.g., rounded v ramified) and readouts like surface area to volume ratio can also provide information about astrocyte morphology.
8. Parameters such as number of neurites, neurite length, branching, and orientation are commonly features of interest. Immunostained neurites can be identified as objects using automated image analysis software, or in some cases it can be beneficial to trace neurites of interest manually (although it is important to note that if this is done using image projections or

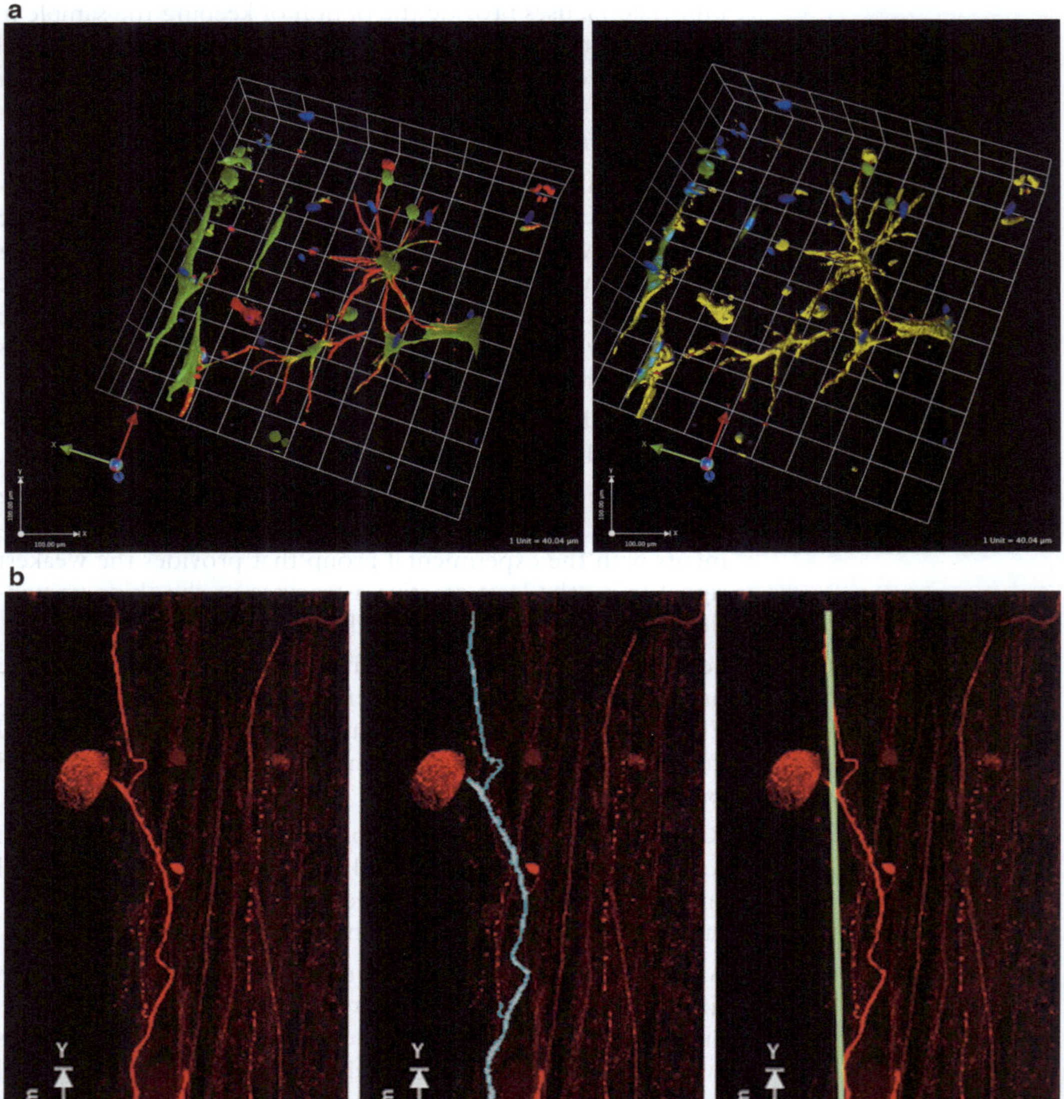

Fig. 2 (**a**) 3D rendered confocal image data (*left*) showing GFP-positive astrocytes (*green*) immunostained to detect GFAP (*red*). An image analysis protocol has been applied to the *red channel* in order to select GFAP-positive objects (shown in *yellow*) for quantification. Nuclei are shown in the *blue channel*. (**b**) Extended focus view of neurons aligned in 3D culture stained with anti-β-tubulin III (*left*) with neurites selected using analysis protocols that identify the neurite structure (*middle*) or overlay a linear representation to aid angle measurement (*right*)

extended focus views then features in the *z*-dimension may not be apparent). A potential limitation with the use of automated thresholding to select neurites is that minor discontinuities in staining can result in single neurites being wrongly identified as multiple separate objects. Some image analysis software will allow similar adjacent objects to be merged automatically, but it is useful to check images carefully to ensure neurites are selected in their entirety. Where the direction of neurite growth is of interest, ensure that the image orientation is appropriate to allow a "bearing" or angle of deviation from one axis of the image to be relevant (Fig. 2b). Where absolute neurite length is required then images must include the cell body and entire neurite, which requires sampling protocols to take this into account.

9. **Steps** 7 and **8** above explain how astrocyte morphology and neuronal growth can each be monitored, and these analyses can be conducted simultaneously in the same fields to explore interactions between the cell types (Fig. 3). For example,

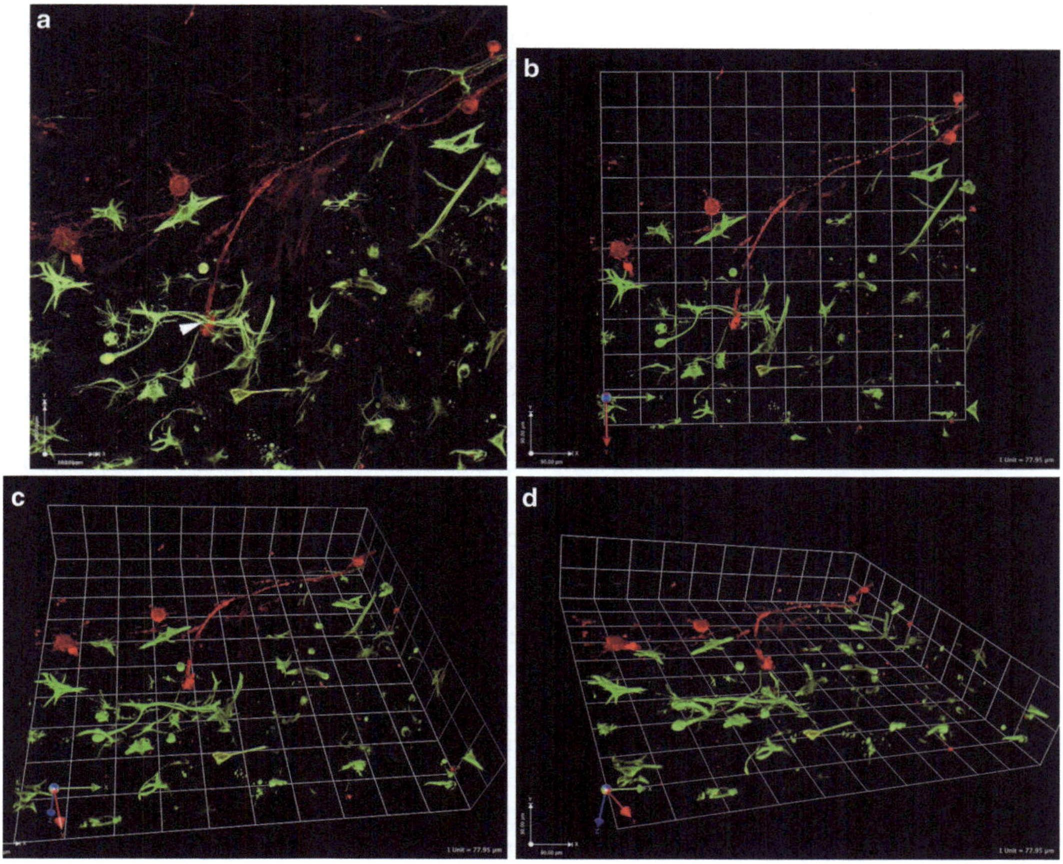

Fig. 3 Neuronal interaction with astrocytes in 3D culture as shown by confocal projection (**a**) followed by tilting of rendered image to reveal spatial distribution of cells in *z*-dimension (**b**–**d**). Note how the apparent close contact between neuron and astrocyte in (**a**) (*arrow*) is revealed to be more distant in three dimensions

neurite length might alter and neurites may turn towards or away from astrocytes exhibiting specific behaviors, so correlating neurite length or orientation of growth with astrocyte morphology or position in particular fields may be of interest [17]. Localized effects of glial alignment on neurite orientation can be detected by either determining the angle of each cell type relative to the sample axis or exploring the angle of deviation of neurite growth from glial cell orientation [24].

4 Notes

1. Commercially available neutralizing reagents provide a rapid and consistent approach but can be costly and may contain unidentified components. An alternative approach is dropwise addition of NaOH until the pH indicator shows the mixture is neutral.
2. Cell seeding density needs to be optimized for the intended application. Higher density 3D cultures approach a more tissue-like structure, but this can make analysis more complex, media must be changed more frequently, and some cells will contract collagen gels at high density which can alter the overall structure of the model. For primary rat astrocytes a cell density of 2–3 million cells per ml of gel is a good compromise, combined with the neurons from 16 dissociated DRGs per ml of gel [8, 15, 17].
3. It is important to ensure proper integration of collagen with the tethering bars prior to setting, and to use an appropriate density of cells to ensure sufficient tension develops within the gels to align the cells [19]. If the astrocytes need to be seeded at a density which is too low for sufficient tension to develop, then stimulating them with TGFβ1 (10 ng/ml for 48 h) or admixing a more contractile cell type (e.g., 5 % fibroblasts) will facilitate contraction and alignment [15].
4. Plastic compression can be used to stabilize cellular collagen gels by removing fluid in order to increase cell and matrix density [25]. Detailed protocols have been published previously [19], and absorbent cell culture inserts are available commercially for this purpose (RAFT plate, TAP Biosystems).
5. The volume of the fluid within the gels must be taken into consideration during incubations with fixative, antibodies and other reagents.
6. Because the *x*- and *y*-dimensions of a microscope field are fixed it is sensible to set a standard *z*-dimension to ensure a consistent volume sample is obtained from each field. The *z*-depth will be determined by the size and shape of gel, location of

cells of interest, and objective lens choice, but should be sufficient to enable the data to be truly 3D rather than a thin slice. The step size (distance between *z*-planes) should be consistent so that *z*-resolution is equivalent between samples—there is an optimal value for this determined by the specifications of the objective lens. The sampling protocol should define the *xy* location of each scan based on landmarks or distances from edges, and also the start and end of the *z*-series as a distance from the top or bottom of the sample. An alternative approach is to sample randomly, but care must be taken to avoid bias in that case (e.g., by the operator not knowing the identity of each sample and selecting fields at low power to avoid observation of particular features of interest).

7. To avoid saturation of fluorescence signals, focus on the brightest apparent feature and adjust the gain until its intensity is within the detectable range. A useful approach is to apply a look-up-table such that gives a specific color to objects that lie beyond the dynamic range. Adjust the gain and offset to give an optimal signal. If this is problematic then the laser power can be adjusted to boost or reduce the fluorescence. For samples with a low signal to noise ratio then averaging can be used to reduce background and increase contrast.
8. Thresholds must be determined to ensure that the volume of staining selected by the image analysis program is reliable. When setting thresholds for inclusion/exclusion it is useful to zoom a region of interest (e.g., a delicate cell process or thin neurite, or conversely a patch of high background staining) and fine tune the settings. Setting exclusion criteria based on minimum object size can be advantageous for samples where debris or background signals are a problem.

References

1. Wang DD, Bordey A (2008) The astrocyte odyssey. Prog Neurobiol 86(4):342–367
2. Silver J, Miller JH (2004) Regeneration beyond the glial scar. Nat Rev Neurosci 5(2): 146–156
3. Lee J, Cuddihy MJ, Kotov NA (2008) Three-dimensional cell culture matrices: state of the art. Tissue Eng Part B Rev 14(1):61–86
4. East E, Phillips JB (2008) Tissue engineered cell culture models for nervous system research. In: Greco GN (ed) Tissue engineering research trends. Nova Science, New York, pp 141–160
5. LaPlaca MC et al (2007) CNS injury biomechanics and experimental models. Prog Brain Res 161:13–26
6. Kofron CM, Fong VJ, Hoffman-Kim D (2009) Neurite outgrowth at the interface of 2D and 3D growth environments. J Neural Eng 6(1):016002
7. Li GN et al (2007) Genomic and morphological changes of neuroblastoma cells in response to three-dimensional matrices. Tissue Eng 13(5):1035–1047
8. East E, Golding JP, Phillips JB (2009) A versatile 3D culture model facilitates monitoring of astrocytes undergoing reactive gliosis. J Tissue Eng Regen Med 3(8):634–646
9. Puschmann TB et al (2013) Bioactive 3D cell culture system minimizes cellular stress and maintains the in vivo-like morphological complexity of astroglial cells. Glia 61(3):432–440

10. Bailey JL et al (2011) In vitro CNS tissue analogues formed by self-organisation of reaggregated post-natal brain tissue. J Neurochem 117(6):1020–1032
11. Dubois-Dauphin ML et al (2010) The long-term survival of in vitro engineered nervous tissue derived from the specific neural differentiation of mouse embryonic stem cells. Biomaterials 31(27):7032–7042
12. Honegger P et al (2011) Preparation, maintenance, and use of serum-free aggregating brain cell cultures. Methods Mol Biol 758:81–97
13. Gil V, del Rio JA (2012) Analysis of axonal growth and cell migration in 3D hydrogel cultures of embryonic mouse CNS tissue. Nat Protoc 7(2):268–280
14. Ge D et al (2013) Culture and differentiation of rat neural stem/progenitor cells in a three-dimensional collagen scaffold. Appl Biochem Biotechnol 170(2):406–419
15. East E et al (2010) Alignment of astrocytes increases neuronal growth in three-dimensional collagen gels and is maintained following plastic compression to form a spinal cord repair conduit. Tissue Eng Part A 16(10):3173–3184
16. Cullen DK, Simon CM, LaPlaca MC (2007) Strain rate-dependent induction of reactive astrogliosis and cell death in three-dimensional neuronal-astrocytic co-cultures. Brain Res 1158: 103–115
17. East E, Golding JP, Phillips JB (2012) Engineering an integrated cellular interface in three-dimensional hydrogel cultures permits monitoring of reciprocal astrocyte and neuronal responses. Tissue Eng Part C Methods 18(7):526–536
18. Cullen DK, Vernekar VN, LaPlaca MC (2011) Trauma-induced plasmalemma disruptions in three-dimensional neural cultures are dependent on strain modality and rate. J Neurotrauma 28(11):2219–2233
19. Phillips JB, Brown R (2011) Micro-structured materials and mechanical cues in 3D collagen gels. Methods Mol Biol 695:183–196
20. East E, Johns N, Georgiou M, Golding JP, Loughlin AJ, Kingham P, Phillips JB (2013) A 3D in vitro model reveals differences in the astrocyte response elicited by potential stem cell therapies for CNS injury. Regen Med 8(6):739–746
21. Gromnicova R, Davies HA, Sreekanthreddy P, Romero IA, Lund T, Roitt IM, Phillips JB, Male DK (2013) Glucose-coated gold nanoparticles transfer across human brain endothelium and enter astrocytes in vitro. PLoS One 8(12):e81043
22. Grinnell F, Petroll WM (2010) Cell motility and mechanics in three-dimensional collagen matrices. Annu Rev Cell Dev Biol 26: 335–361
23. Sundararaghavan HG et al (2009) Neurite growth in 3D collagen gels with gradients of mechanical properties. Biotechnol Bioeng 102(2):632–643
24. Georgiou M et al (2013) Engineered neural tissue for peripheral nerve repair. Biomaterials 34(30):7335–7343
25. Brown RA et al (2005) Ultrarapid engineering of biomimetic materials and tissues: fabrication of nano- and microstructures by plastic compression. Adv Funct Mater 15: 1762–1770

Part IV

Axon Injury and Regeneration In Vivo

Chapter 10

Targeting Inhibitory Chondroitin Sulphate Proteoglycans to Promote Plasticity After Injury

Jessica C.F. Kwok, Janosch P. Heller, Rong-Rong Zhao, and James W. Fawcett

Abstract

Chondroitin sulphate proteoglycans (CSPGs) are one of the major families of inhibitory extracellular matrix molecules in the central nervous system. The expression of various CSPGs is strong during early nervous system development; however, it is downregulated during maturation and up-regulated again after nervous system injury. In vivo injection of an enzyme called chondroitinase ABC, which removes the inhibitory chondroitin sulphate chains on the CSPGs, in the injured area promotes both the regeneration and plasticity of the neurons. Here, we describe the method of in vivo injection of the chondroitinase ABC into the cortex of adult rat brain and the histochemical method to assess the successfulness of the digestion.

Key words Chondroitin sulphate proteoglycans, Perineuronal nets, Extracellular matrix, Chondroitinase ABC, Plasticity, Regeneration, Central nervous system, Spinal cord injury

1 Introduction

The extracellular matrix (ECM) in animals is mainly composed of proteins (mainly fibrous proteins such as collagen, fibronectin, and laminin) and glycosaminoglycans (GAGs)/proteoglycans (PGs). The general role of the ECM is to provide a scaffold for the structure of the surrounding tissue. However, the ECM also bears signalling properties and is involved in tissue development, maturation, differentiation, cell adhesion and migration, cell survival, and tissue homeostasis [1]. In the central nervous system (CNS), ECM has additional functions in regulating neurite extension together with adhesion and migration of neurons. One major family of the ECM molecules which controls the abovementioned processes in the CNS is chondroitin sulphate proteoglycan (CSPG) [2, 3]. During early development, CSPGs are involved in the shaping of the neuronal circuitry either by acting directly as an inhibitory molecule or through binding to different growth factors, therefore limiting their availability [4]. Apart from controlling circuitry formation

Andrew J. Murray (ed.), *Axon Growth and Regeneration: Methods and Protocols*, Methods in Molecular Biology, vol. 1162, DOI 10.1007/978-1-4939-0777-9_10,

during development, CSPGs are also strongly up-regulated after spinal cord injury [5, 6]. After CNS injury, adult neurons show very limited potential to regenerate [7–9]. This regenerative failure is due to a loss of growth-promoting trophic factors and substrate molecules [10, 11] and an up-regulation of growth inhibitory molecules such as CSPGs, myelin-associated molecules such as Nogo-A, and myelin-associated glycoprotein [5, 8]. Digestion of chondroitin sulphate (CS) on the CSPGs using an enzyme called chondroitinase ABC (ChABC) successfully removes the CSPG inhibition [5, 12]. Animals treated with the ChABC demonstrate significant improvement in both structural and functional recovery [13, 14]. Recently, CSPGs have also been shown to limit plasticity in the adult nervous system [15–17]. Apart from being found in the loose ECM, CSPGs are also found surrounding the soma and proximal dendrites in a structure called perineuronal net (PNN). PNN is a layer of ECM which wraps on the neuronal surface, where synaptic connections are made. PNNs are formed at the end of the critical period when the neuronal system lost the majority of its plastic ability [17, 18]. After CNS injury, a significant proportion of functional recovery is achieved through axonal sprouting and dendritic remodelling [19–22]. However, these pro-recovery events are hampered by the up-regulated CSPGs around the injury sites and also on the surface of the spared neurons as PNNs. The axons of the injured neurons form dystrophic growth cones upon interaction with the CSPGs [23]. Once again, application of ChABC to the relevant brain regions is able to remove this plasticity brake and enhance recovery [16, 24, 25]. Activation of plasticity is not just improving recovery after CNS injury; recent studies show that removing CSs in the PNNs may also benefit other neurological conditions such as dementia, memory, and learning [26, 27]. Thus, ChABC seems to be a useful tool in enhancing regeneration and for studying plasticity in different animal models. Here, we describe the method of in vivo digestion of CSs in adult rat brain by ChABC injection and the histochemical techniques used to confirm a successful enzymatic digestion.

2 Materials

2.1 Chemicals and Solutions

1. Operating isoflurane (e.g., Abbott, 1–2 % in a mixture of 25 % nitric oxide and 50 % oxygen).
2. Absolute ethanol: Dilute it with different volume of deionised water (D-H_2O) for different experimental steps (such as 70 % for disinfection).
3. Analgesics (e.g., Carprieve, Norbrook, at 5 mg/kg body weight) and antibiotics (e.g., Terramycin, Pfizer, 60 mg/kg).
4. Betadine.
5. Lubricant eye ointment (e.g., Allergan).

6. Sterile saline (e.g., Aqupharm): Warm up to 37 °C before use.
7. ChABC (protease-free form isolated from *Proteus vulgaris*, Seikagaku, Japan): Dissolve in 0.1 M sterile PBS to a final concentration of 100 U/ml. Store the aliquots at −20 °C. On the day of injection, thaw the aliquot on ice and keep the solution in the icebox until injection.
8. Pentobarbitone.

2.2 Animals and Stereotaxic Surgery

1. Animals: 3-month-old adult male Sprague Dawley rats from Charles River (UK) are used for the experiment (*see* **Note 1**). The animals are housed in groups of up to four rats under a 12-h/12-h light/dark cycle (lights on from 6 a.m. to 6 p.m.) with room temperature (rtp) at 21 °C. Food and water are given *ad libitum*. All procedures are performed in accordance with the UK Animals (Scientific Procedures) Act (1986).
2. Brain atlas for stereotaxic coordinates [28].
3. Stereotaxic apparatus: Stereotaxic frame for small animals (David Kopf Instruments, UK).
4. Laboratory scale.
5. Electric hair shaver.
6. Temperature-controlled heating mat.
7. Neurosurgery materials: Sterile cotton swabs (Johnson & Johnson, autoclaved), surgical forceps [e.g., Dumont #5 forceps from Fine Science Tool (FST)], scalpel and scalpel blades, blunt dissection scissors (FST), skin retractor (FST) or "surgical hooks" (made from clipped and bent needles), surgical drill and small bits, Hamilton syringe (Hamilton 65459-01), vertical micro-drive pump with syringe holder which can be fixed onto the stereotaxic frame manipulator (e.g., UMP2 pump, World Precision Instrument), 4-0 suture needles and needle holder (FST), 27-G needles, 1- and 2.5-mL disposable syringes (for subcutaneous injection of the analgesic and antibiotic solution), ligature scissors for removing sutures (FST), surgical microscope.

2.3 Perfusion of Animals

1. Buffered saline pre-wash (pH7.4): Dissolve 1.78 g disodium hydrogen orthophosphate $\cdot 2H_2O$, 9 g sodium chloride in 1 l of D-H_2O (pH 7.4).
2. 4 % paraformaldehyde (PFA, pH7.4): Dissolve 8 g of PFA in 200 ml of D-H_2O.
3. 1× phosphate-buffered saline (PBS): Dissolve 4 g of sodium chloride, 0.1 g of potassium chloride, 0.12 g of potassium dihydrogen phosphate, and 0.72 g of disodium hydrogen phosphate in 500 ml of D-H_2O, adjust the pH to 7.4.
4. 30 % (w/v) sucrose solution: Dissolve 30 g of sucrose in 100 ml of 1× PBS.

2.4 Tissue Sectioning and Histochemistry

1. 24-well plates.
2. Fine-hair paintbrush.
3. 1× PBS with 0.05 % sodium azide (PBS-azide): Dissolve 0.05 g of sodium azide in 100 ml of 1× PBS (*see* **Note 2**).
4. TS-PBS: Prepare a 10 ml solution by mixing 0.3 ml of normal goat serum, 0.03 ml of Triton X-100, and 9.67 ml of 1× PBS.
5. PBS-T: Prepare 100 ml of the solution by mixing 0.3 ml of Triton X-100 with 99.7 ml of 1× PBS.
6. Biotinylated *Wisteria floribunda* agglutinin (Sigma): Reconstitute the lectin into 1 mg/ml solution in 1× PBS. Aliquot and store at −20 °C.
7. ABC solution: Mix one volume of solution A with one volume of solution B in 48 volume of 1× PBS from using the reagents in the VECTASTAIN ABC kit "Standard" (Vector Labs).
8. Tris buffered non-saline, pH 7.4 (TNS): Dissolve 6 g of Trizma base in 1 l of D-H_2O. Adjust the pH to 7.4.
9. 3,3′-Diaminobenzidine (DAB) solution (Sigma): Dissolve the tablet of DAB substrate and urea/hydrogen peroxide in 5 ml of D-H_2O. Keep the solution in the dark and use within 1 h (*see* **Note 3**).
10. Gelatine-coated slides: Slowly dissolve 0.5 g of gelatin in 500 ml of D-H_2O at 56 °C in 5–10 min for 0.1 % (w/v) gelatin solution. Wash the glass slides in 0.1 M sodium hydroxide for 15 min, then rinse with running tap water, and drain. Put the slides into slide holders and dip into acetone 3–4 times (perform this step in a fume hood). Air-dry the slides for a few minutes in the fume hood. Then dip the slides for 3–4 times in the 0.1 % gelatin solution. Drain the excess solution, and leave the slides in a 37 °C oven overnight. Box up and store the slides in a dry place.
11. Xylene.
12. DPX mounting medium.
13. Microscope.

3 Methods

3.1 Preparation for the Stereotaxic Frame and Injection Coordinates

1. The coordinates are given as three-dimensional (*x*, *y*, and *z*) distances (in mm) from the bregma, which is the intersection of the coronal and sagittal sutures on the surface of the skull. The *x* plane represents the medial-lateral (ML) distance, the *y* plane represents the anterior–posterior (AP) distance (where a prefix "+" indicates the anterior end and "−"as the posterior end), and the *z* plane represents the dorsal–ventral (DV) distance (where a prefix "+" indicates the ventral side) from the bregma.

2. Injection coordinates (mm) for this experiment in reference to the bregma are:
   - First injection: AP −0.7, ML −1.5, DV +1.5.
   - Second injection: AP −0.7, ML −1.4, DV +1.5 (*see* **Notes 4** and **5**).

3.2 Preparation of Surgical Tools and Operation Area

1. Before the start of the surgery, ensure that all surgical tools are clean and sterile (either by autoclaving or by immersion in disinfectant) and are ready to use for the operation. Clean the operating area with 70 % ethanol.
2. Ensure that the respiration and scavenging systems are working properly and the oxygen level is sufficient to support the entire surgery.
3. Ensure that the operating microscope and the lights are working properly.
4. Connect the drill and the micro-drive pump to the power supply, and check the devices to ensure proper functions. Set up the parameters of the pump (speed: 100 nl/min, volume: 1,000 nl).
5. Fix the drill vertically onto the stereotaxic frame probe holder, which is then secured to the *x*-axis of the stereotaxic frame manipulator through the v-shape mounting clamp. Lift the drill up to allow sufficient space for placing the rat.

3.3 Preparation of Animal and Neurosurgery

1. All surgical procedures are performed under inhalation anesthetic with isoflurane.
2. Inject analgesics subcutaneously 2 h before surgery (*see* **Note 6**).
3. Weigh the rat before anesthesia, and check that the animal is in a good health condition (*see* **Note 7**).
4. Anesthetize the rat in an induction box, observe closely, and wait until the rat stops moving and the breathing rate is stable at a reduced rate.
5. Move the rat to the shaving area, and place the nose at the opening of the isoflurane tube (*see* **Note 8**).
6. Shave the skull gently, blow away the loose hairs, and clean the skin with Betadine.
7. Move the rat to the stereotaxic surgical stage, and switch on the isoflurane.
8. The body temperature should be maintained at 37 °C with a heating pad during the entire surgery.
9. Apply an eye ointment including vitamin A to protect the eyes.
10. Pinch the feet with nails to make sure that the rat reaches deep anesthesia and no reflex is shown. Otherwise wait for longer time until deep anesthesia is reached.

11. To place the rat onto the stereotaxic frame fix one ear bar in the apparatus, and gently position the rat's head so that its ear canal is securely placed onto the ear bar. Keep the head in place, and slowly push the second ear bar to the other ear canal. Stop immediately when feeling that a moderate pressure is encountered. Screw tight the second ear bar to fix the head in position. The head should be horizontal to the ear bars and should not move laterally.
12. Secure the rat onto an incisor (front teeth) adapter: Gently open the lower jaw, slowly move the incisor adapter into the mouth, and fit the incisors at the opening of the adapter. Pull back the adaptor slightly, and fix it in place. Then lower the nose clamp onto the nose only with a low pressure (*see* **Note 9**).
13. Check that the tongue is in a normal position; otherwise, pull it out gently with small blunt forceps to prevent airway blockage (*see* **Note 10**).
14. Adjust the surgical microscope, and use low magnification to obtain a clear and complete view of the top of the skull.
15. Next, make a midline skin incision with scalpel blade 10 starting from caudal to the midpoint between the eyes.
16. Retract the skin sideways with surgical hooks or skin retractor from the incision line (*see* **Note 11**).
17. Measure the z coordinates of the bregma and the lambda, and adjust the head level using the incision adaptor so that the z coordinates are equal at the two points.
18. Measure the x and y coordinates of the bregma, calculate the target coordinates for injection, and adjust the manipulator to the target position accordingly.
19. Lower the drill bit until it is almost touching the skull. Increase the microscope magnification, and adjust the focus to have a clear view of the target area.
20. Switch on the drill, and carefully drill a hole in the skull. Stop drilling when the blood vessels on the dura become visible and only a very thin layer of the skull is left. Remove the thin skull layer with a pair of fine forceps (*see* **Note 12**).
21. Repeat the above procedure for the second coordinate.
22. Prepare for the injection: Remove the probe holder together with the drill from the manipulator and secure the micro-drive pump onto the x-axis via the mounting clamp.
23. Load the ChABC solution into the Hamilton syringe until the reading reaches 2 μl for each animal (*see* **Note 13**).
24. Carefully fix the syringe onto the pump, make sure that the syringe is secured vertically, and the scale can be seen conveniently. Check that the pump parameters are set correctly at 100 nl/min or 6 μl/h for 10 min.

25. Pierce through the dura with a sharp needle tip (*see* **Note 14**).
26. Bring the Hamilton needle to the correct *x* and y coordinates, and lower it until it touches the dura.
27. Slowly lower the needle to the desired *z* coordinate, and start the injection. Allow the pump to run for 10 min (*see* **Note 15**).
28. Remove the needle slowly to avoid a backflow of the solution.
29. Repeat the injection procedure for the second injection.
30. Clean the opening with moist cotton swabs, and remove all debris including loose hair or small bone fragments.
31. Suture the skin on the skull, and inject antibiotics subcutaneously. Inject warm saline (37 °C) subcutaneously to avoid dehydration of the animal after surgery.

3.4 Recovery from Anesthesia

1. Keep the rat at a warm place (in a cage with blanket or in a recovery incubator) until fully recovered. When fully recovered, return the rat to a clean cage (*see* **Note 16**).
2. Check the animals every day in the following week for signs of stress including wound inflammation, pain, wound scratching, suture opening, and weight loss (*see* **Note 17**).
3. Remove the suture 7 days after operation under brief inhalation anesthesia.

3.5 Perfusion of Rat

1. The post-op rat is terminally anesthetized with pentobarbitone, followed by perfusion with 200 ml of cold PBS pre-wash through the heart and then 4 % PFA (*see* **Note 18**).
2. The brain is removed and post-fixed for 24 h at 4 °C.
3. The brain is then transferred to cold 30 % sucrose for cyroprotection at least overnight at 4 °C before sectioning.

3.6 Tissue Sectioning and Histochemistry (See Note 19)

1. Take out the brain from the sucrose solution, and briefly blot it dry on a 3 mm filter paper.
2. Freeze down the brain on a sledge microtome platform, and section the brain into 35 μm thick coronal or parasagittal sections in PBS-azide (*see* **Note 20**).
3. To check if the ChABC has successfully digested the CS on the PNNs in the brain, we use a lectin called *Wisteria floribunda agglutinin* (biotinylated, bio-WFA) which labels the CS on the CSPGs (*see* **Note 21**).
4. The brain sections are washed in 1 ml of 1× PBS for 5 min, three times at rtp.
5. Block the sections with TS-PBS for 1 h at rtp.
6. Remove the blocking solution, add 20 μg/ml of bio-WFA (in TS-PBS), and incubate overnight at 4 °C.
7. Remove the lectin solution, and wash the sections in PBS-T for 15 min, three times at rtp.

8. While washing the sections, prepare the ABC solution (VECTASTAIN ABC kit "Standard") by mixing 1:1:48 of reagent A:reagent B:1× PBS, and let it stand for at least 1/2 h at rtp for the formation of ABC complex before use (*see* **Note 22**).
9. Add the premixed ABC solution to the sections, and incubate for 1 h at rtp.
10. Rinse the sections with TNS, three times for 15 min each.
11. Prepare the DAB solution (*see* **Note 23**).
12. Incubate the sections in DAB solution, and monitor the color intensity upon the addition of DAB substrate. Stop the reaction by removing the DAB solution, and wash the sections twice with TNS for 5 min.
13. Put the sections in a big Petri dish, and mount the sections onto gelatin-coated slides. Air-dry the sections for 1 h.
14. Dehydrate the sections through a series of ethanol solution of increasing percentage (50, 70, 90 % and two times of 100 % EtOH), 5 min per solution. The slides are then incubated twice in xylene for 5 min at rtp. The slides are coverslipped with DPX Mounting Medium (*see* **Note 24**).
15. The slides are air-dried overnight at rtp in a fume hood before documentation using a bright-field microscope (Fig. 1).

4 Notes

1. Other rat strains and mice can also be used for the injection.
2. Sodium azide is a toxic chemical. Please follow the supplier's instruction and local regulation for proper handling and disposal.
3. DAB is a strong mutagen. Please wear gloves when handling the chemical. All solutions and consumables in contact with the DAB have to be disposed according to the supplier's instruction and local regulations.
4. To obtain targeting coordinates for a specific injection region, subtract the atlas coordinates from the position of the animal's bregma in the stereotaxic apparatus. For example, if the bregma's coordinates are "AP 60, ML 15, DV 30," then the desired injection coordinates are "AP 60.7, ML 16.5, DV 28.5" for the first injection and "AP 60.7, ML 16.4, DV 28.5" for the second injection.
5. The coordinates used in this protocol correspond to the sensorimotor cortex of the hindlimb in an adult rat. We chose this area because hindlimb functions are mostly compromised after lesion in various spinal cord injury models. Removing inhibitory PNNs using ChABC opens up a window of plasticity in the selected region and may enhance the functional recovery of the

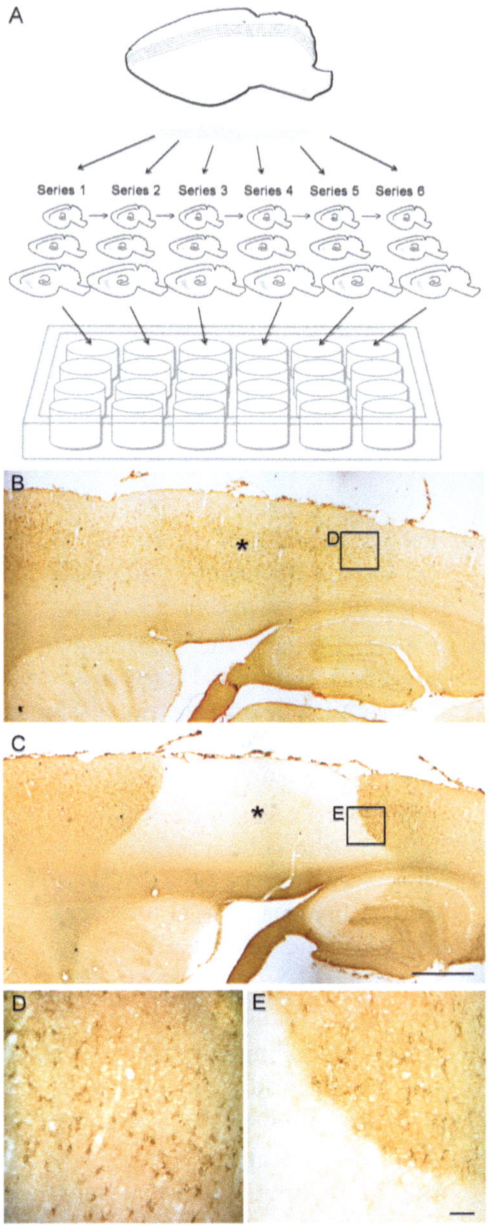

Fig. 1 Section collection (**a**) and staining of chondroitin sulphates with bio-WFA (*brown*) in the adult cortex 24 h after the injection of PBS (**b**, **d**) or ChABC (**c**, **e**). (**a**) The brain sections should be collected as a six-section series to ensure that every stack of sections covers all levels in the brain. (**b**) and (**d**) PBS injection does not remove the WFA-positive signal in the adult cortex. Subpopulations of neurons are clearly wrapped by perineuronal nets under high magnification in (**d**). (**c**) and (**e**) 24 h after the injection of the ChABC, a clear area devoid of *brown* WFA signal (both in the extracellular matrix and on the cell surface) is observed around the injected area. High-magnification image in (**e**) shows very faint remnants of perineuronal staining on neurons present in the ChABC area, while in the area outside the digestion zone, a very clear perineuronal staining is found on the neurons. The *asterisks* mark the center of the injection sites. Scale bar in (**b**) and (**c**) is 1 mm (**d**) and (**e**) is 100 μm

corresponding body part in response to different treatments and rehabilitation program.

6. The analgesic is injected subcutaneously 2 h before the surgery to ensure that the optimal level of analgesic effect is reached at the point of the surgery. It can also be injected before the start of the surgery.
7. A good health condition is important for a speedy recovery and to avoid unnecessary complications. If an animal is ill or under stress before the surgery, either postpone the surgery or choose a healthier substitute for the operation.
8. It is important to make sure that the tube is fitted properly to the nose of the animal. Any leakage from the tube will reduce the efficiency of anesthesia and the leaked isoflurane will be released to the operation area. This poses potential harm to the researcher.
9. Proper securing of the animal to the stereotaxic frame is critical for the operation. For detailed descriptions of the procedure, please refer to ref. [29].
10. Closely monitor the breathing of the animal throughout the surgery. If the animal starts to show abnormal, slow, or labored breathing, release the animal from the stereotaxic apparatus. Provide air without isoflurane until the breathing recovers to a normal pattern. Check that the oxygen is at a sufficient level.
11. In order to reveal a clear view of the skull, scientist can gently separate the subcutaneous tissue from the skull using blunt dissection scissors. One can further clean the connective tissue on the bregma and lambda areas using sterile surgical blade. To avoid the skin tissue from drying out, apply several drops of sterile PBS to the tissue along the incision line using wet cotton swabs.
12. To avoid drilling through the whole skull and damaging the cortex, make sure that you only remove a small depth of the skull each time and stop frequently to check for the depth.
13. The total injection volume is 2 μl (1 μl for each injection site). However, when loading the solution for the injection setup, the dead volume of the syringe should also be included (usually ~1 μl).
14. Hold the needle at a flat angle to avoid damaging the brain tissue.
15. When the injection is finished, leave the needle in the injection site for another 3 min before retraction. This is to ensure that the injected solution is fully absorbed into the surrounding tissues.
16. It is helpful to place a small dish of soaked food pellets in the cage for easy food access after surgery.
17. The image of the histology performed in this chapter is on an adult rat brain collected at 24 h after ChABC injection.

Our previous study shows that a single ChABC injection in the cortex would effectively keep the CS at a low level for at least 10 days after injection [30].

18. A good perfusion is important for minimizing background staining from histochemistry and to maintain a nice tissue fixation. Perfusion with PBS pre-wash is to rinse out blood cells in the blood vessels which tend to give strong background after staining.
19. Histology is only one of the many ways in determining the success of ChABC digestion. One can also assay for the residual ChABC activity recovered from the injected tissues if necessary [31, 32].
20. The sections are usually collected as six-series stacks in the first row of a 24-well plate containing PBS-azide (Fig. 1). This is to ensure that each well would have sections covering the different levels from the brain. The sections can be stored at 4 °C for at least 3 months.
21. WFA is a lectin which demonstrates a very strong and specific binding affinity to CSs. One can also use other antibodies to detect CS in the area, such as clone CS-56 for CSs (Sigma) and clone 1 F6 for neurocan (DSHB).
22. If fluorescence detection is preferred, use fluorophore-conjugated streptavidin instead. After incubation for 2 h at rtp, rinse the sections in 1× PBS three times for 15 min each, before mounting the section with anti-fading mounting medium.
23. DAB is a toxic and carcinogenic chemical. Please follow the provider's instruction and local laboratory practice and regulation for proper handling and disposal.
24. Xylene and DPX are very volatile solution and highly flammable. Please carry out these steps in a fume hood. Xylene is a strong organic solvent, so use anticorrosive plasticwares or glass containers for all the procedures.

References

1. Novak U, Kaye AH (2000) Extracellular matrix and the brain: components and function. J Clin Neurosci 7:280–290
2. Rhodes KE, Fawcett JW (2004) Chondroitin sulphate proteoglycans: preventing plasticity or protecting the CNS? J Anat 204:33–48
3. Yamaguchi Y (2000) Lecticans: organizers of the brain extracellular matrix. Cell Mol Life Sci 57:276–289
4. Kwok JC, Dick G, Wang D, Fawcett JW (2011) Extracellular matrix and perineuronal nets in CNS repair. Dev Neurobiol 71:1073–1089
5. Lemons ML, Howland DR, Anderson DK (1999) Chondroitin sulfate proteoglycan immunoreactivity increases following spinal cord injury and transplantation. Exp Neurol 160:51–65
6. Zuo J, Neubauer D, Dyess K, Ferguson TA, Muir D (1998) Degradation of chondroitin sulfate proteoglycan enhances the neurite-promoting potential of spinal cord tissue. Exp Neurol 154:654–662
7. Bravin M, Savio T, Strata P, Rossi F (1997) Olivocerebellar axon regeneration and target reinnervation following dissociated Schwann cell grafts in surgically injured cerebella of adult rats. Eur J Neurosci 9: 2634–2649

8. Chen MS, Huber AB, van der Haar ME, Frank M, Schnell L, Spillmann AA, Christ F, Schwab ME (2000) Nogo-A is a myelin-associated neurite outgrowth inhibitor and an antigen for monoclonal antibody IN-1. Nature 403: 434–439
9. Morrow DR, Campbell G, Lieberman AR, Anderson PN (1993) Differential regenerative growth of CNS axons into tibial and peroneal nerve grafts in the thalamus of adult rats. Exp Neurol 120:60–69
10. Kawaja MD, Gage FH (1991) Reactive astrocytes are substrates for the growth of adult CNS axons in the presence of elevated levels of nerve growth factor. Neuron 7:1019–1030
11. Smith GM, Jacobberger JW, Miller RH (1993) Modulation of adhesion molecule expression on rat cortical astrocytes during maturation. J Neurochem 60:1453–1466
12. Jones LL, Margolis RU, Tuszynski MH (2003) The chondroitin sulfate proteoglycans neurocan, brevican, phosphacan, and versican are differentially regulated following spinal cord injury. Exp Neurol 182:399–411
13. Yick LW, Wu W, So KF, Yip HK, Shum DK (2000) Chondroitinase ABC promotes axonal regeneration of Clarke's neurons after spinal cord injury. Neuroreport 11:1063–1067
14. Bradbury EJ, Moon LD, Popat RJ, King VR, Bennett GS, Patel PN, Fawcett JW, McMahon SB (2002) Chondroitinase ABC promotes functional recovery after spinal cord injury. Nature 416:636–640
15. Galtrey CM, Asher RA, Nothias F, Fawcett JW (2007) Promoting plasticity in the spinal cord with chondroitinase improves functional recovery after peripheral nerve repair. Brain 130:926–939
16. Barritt AW, Davies M, Marchand F, Hartley R, Grist J, Yip P, McMahon SB, Bradbury EJ (2006) Chondroitinase ABC promotes sprouting of intact and injured spinal systems after spinal cord injury. J Neurosci 26: 10856–10867
17. Pizzorusso T, Medini P, Berardi N, Chierzi S, Fawcett JW, Maffei L (2002) Reactivation of ocular dominance plasticity in the adult visual cortex. Science 298:1248–1251
18. McRae PA, Rocco MM, Kelly G, Brumberg JC, Matthews RT (2006) Sensory deprivation alters aggrecan and perineuronal net expression in the mouse barrel cortex. J Neurosci 27:5405–5413
19. Fawcett JW (2009) Recovery from spinal cord injury: regeneration, plasticity and rehabilitation. Brain 132:1417–1418
20. Fawcett JW, Curt A (2009) Damage control in the nervous system: rehabilitation in a plastic environment. Nat Med 15:735–736
21. Garcia-Alias G, Barkhuysen S, Buckle M, Fawcett JW (2009) Chondroitinase ABC treatment opens a window of opportunity for task-specific rehabilitation. Nat Neurosci 12:1145–1151
22. Girgis J, Merrett D, Kirkland S, Metz GA, Verge V, Fouad K (2007) Reaching training in rats with spinal cord injury promotes plasticity and task specific recovery. Brain 130: 2993–3003
23. Tom VJ, Steinmetz MP, Miller JH, Doller CM, Silver J (2004) Studies on the development and behavior of the dystrophic growth cone, the hallmark of regeneration failure, in an in vitro model of the glial scar and after spinal cord injury. J Neurosci 24:6531–6539
24. Cafferty WB, Bradbury EJ, Lidierth M, Jones M, Duffy PJ, Pezet S, McMahon SB (2008) Chondroitinase ABC-mediated plasticity of spinal sensory function. J Neurosci 28: 11998–12009
25. Fawcett J (2009) Molecular control of brain plasticity and repair. Prog Brain Res 175: 501–509
26. Romberg C, Yang S, Melani R, Andrews MR, Horner AE, Spillantini MG, Bussey TJ, Fawcett JW, Pizzorusso T, Saksida LM (2013) Depletion of perineuronal nets enhances recognition memory and long-term depression in the perirhinal cortex. J Neurosci 33: 7057–7065
27. Gogolla N, Caroni P, Lüthi A, Herry C (2009) Perineuronal nets protect fear memories from erasure. Science 325:1258–1261
28. Paxinos G, Watson C (1986) The rat brain in stereotaxic coordinates, 2nd edn. Academic, San Diego
29. Cetin A, Komai S, Eliava M, Seeburg PH, Osten P (2007) Stereotaxic gene delivery in the rodent brain. Nat Protoc 1:3166–3173
30. Lin R, Kwok JC, Crespo D, Fawcett JW (2008) Chondroitinase ABC has a long-lasting effect on chondroitin sulphate glycosaminoglycan content in the injured rat brain. J Neurochem 104:400–408
31. Hyatt AJ, Wang D, Kwok JC, Fawcett JW, Martin KR (2010) Controlled release of chondroitinase ABC from fibrin gel reduces the level of inhibitory glycosaminoglycan chains in lesioned spinal cord. J Control Release 147: 24–29
32. Zhao RR, Muir EM, Alves JN, Rickman H, Allan AY, Kwok JC, Roet KC, Verhaagen J, Schneider BL, Bensadoun JC, Ahmed SG, Yáñez-Muñoz RJ, Keynes RJ, Fawcett JW, Rogers JH (2011) Lentiviral vectors express chondroitinase ABC in cortical projections and promote sprouting of injured corticospinal axons. J Neurosci Methods 201:228–238

Chapter 11

Polymeric Biomaterials for Nerve Regeneration: Fabrication and Implantation of a Biodegradable Nerve Guide

Wesley N. Sivak, Jacqueline M. Bliley, and Kacey G. Marra

Abstract

Optimizing the quantity, quality, and speed of axon regeneration is important in maximizing functional outcomes following peripheral nerve injury. When severed, injured nerves must be able to regenerate and reconnect to the structures they previously controlled within 12–18 months before sensation and motion are permanently lost. Nerve sprouts from the proximal stump will spontaneously migrate toward the distal stump in the event of a nerve transection. However, surgical intervention remains necessary to repair transection injuries. Regeneration becomes particularly troublesome with large gaps, where autologous nerve grafts or nerve guides are used to repair transected nerves. Nerve conduits function as therapeutic adjuncts, guiding axonal regeneration across gap defects. Despite the availability of several FDA-approved nerve conduits, functional outcomes following their use remain less than optimal. Much work has been focused on developing nerve conduits to improve peripheral nerve repair outcomes. This chapter describes fabrication of a poly(caprolactone) nerve guide and demonstrates its use in a rat sciatic nerve model.

Key words Nerve guide, Axonal regeneration, Poly(caprolactone), Drug delivery, Double-walled microspheres, Biomaterial, Biodegradable, Sciatic nerve

1 Introduction

Damage to the peripheral nervous system is surprisingly common. Peripheral nerve injury is thought to occur in 2–3 % of all major traumas and also regrettably results as a complication of surgery [1]. Although complete recovery of nerve function occurs in many mild injuries, outcomes are often unsatisfactory following complete transection. Once cut, nerves must be able to slowly grow, find, and reconnect to the structures they control. When an axon is disconnected from its cell body by injury, its distal segment gradually degenerates through a process called Wallerian degeneration, eventually disappearing [2]. If regrowth and reinnervation do not occur within 12–18 months, sensation and/or motion are permanently lost and there is little chance for functional recovery [3, 4].

Andrew J. Murray (ed.), *Axon Growth and Regeneration: Methods and Protocols*, Methods in Molecular Biology, vol. 1162,
DOI 10.1007/978-1-4939-0777-9_11,

Maximizing sensation and motion in an injured or a transplanted limb appears to lie within optimizing the quantity, quality, and speed of nerve regeneration following repair.

With respect to repair of peripheral nerve injury, end-to-end anastomosis of severed nerve stumps is the preferred method [5]. Unfortunately, inelastic recoil occurs within severed nerve stumps and in the setting of traumatic injury this often results in irreparable, segmental loss of nerve [1]. Nerve autografting is the most commonly used surgical technique for repair of peripheral nerve gaps and remains the clinical gold standard [6]. The graft is typically a sensory nerve harvested from a distant site on the body. The success of autologous nerve grafting arises from the presence of Schwann cells and basal lamina endoneural tubes, which provide neurotrophic factors and surface adhesion molecules to regenerating axons [7]. Some studies have reported complete restoration of sensation after autografting, but the recovery rate of motor function remains less than 40 % [8]. Autografting also depends upon donor nerve availability and compatibility in terms of tissue size and structure. Major disadvantages include limited graft material, loss of function at the donor site, and donor site morbidity (e.g., scarring and painful neuroma formation). In addition, the individual endoneural tubes cannot be exactly approximated in size or number, leading to mismatching of regenerating axons and nonspecific, incomplete reinnervation of the target tissues. Allografts and xenografts have also been studied, but undesired immune responses or requirement for preprocessing to remove immunogenic material limits their effectiveness [2].

A nerve conduit made of biodegradable materials is a promising strategy for spanning gap defects and promoting long-term nerve recovery—a fact confirmed both experimentally and clinically. A well-engineered biodegradable guide initially provides an adequate scaffold for regeneration while retaining the necessary mechanical stiffness to withstand the pressure from the surrounding tissues—all without causing damage to the nerve stumps and adjacent structures. An ideal guide is also semipermeable and allows for controlled exchange between intra- and extra-channel environments; the ability to retain neurotrophic factors within the lumen has been identified as a key factor in increasing nerve growth response [9]. Nerve guides should also aid in excluding immune and inflammatory cells that can promote fibrosis at the repair site. Lastly, the structure should degrade at an appropriate speed as the nerve regenerates. Too quickly, and the walls of the lumen can collapse on the growing nerve or fibrous tissue can infiltrate and disrupt regeneration. In order to be satisfactory for clinical use, an artificial nerve guide must ensure satisfactory nerve repair at least to a level equivalent with autogenous grafts. Several biodegradable nerve guide implants made of collagen, poly(glycolic) acid, and poly(lactide-*co*-caprolactone) have been approved by the FDA for human application and are in clinical use, but indications are limited to gaps less than three centimeters [1]. Silk-based guides are a relatively new alternative and have shown promise in experimental models, but are still in development for clinical use [10].

Many studies have shown that it is possible to improve upon the performance of nerve conduits by enabling them to deliver growth factors and other cellular components thought to be important for nerve growth [11, 12]. Biodegradable grafts have a significant advantage in that, as they degrade, they can be designed to release trophic factors from within or adsorbed to the material. Entrapment of a promoting factor within the walls of the nerve conduit and its release during degradation enable prolonged bioavailability of the factor during the regeneration process. Various neurotrophic factors such as fibroblast growth factor (FGF), neuron growth factor (NGF), glial cell line-derived neurotrophic factor (GDNF), ciliary neurotrophic factor (CNF), and vascular endothelia growth factor (VEGF) have improved device performance when loaded into nerve conduits [6, 13, 14]. A recent study shows that differential gradients of GDNF are able to further drive axonal growth down silk conduits in experimental models of peripheral nerve repair [15].

In this chapter, we present in detail a protocol for fabricating a nerve guide and demonstrate how to test it in an animal model in the laboratory. Instructions are provided for the preparation and sterilization of a poly(caprolactone) (PCL) nerve guide. Further demonstration of the use of this system in the rat sciatic nerve transection model follows. This well-established model can easily be employed for investigation of conduit-based nerve repair strategies. The fabrication techniques described within can easily be adapted to include a variety of drug delivery strategies focused on enhancing the response of the regenerating nerve. The surgical model presented here is well accepted in the literature for preclinical evaluation; the results are easily reproducible following a slight learning curve. The techniques outlined within this chapter should familiarize the reader with how biomaterials can be used to augment the peripheral nerve repair response.

2 Materials

2.1 Chemicals, Solutions, and Laboratory Equipment

1. Nerve guide fabrication:
   (a) Poly(vinyl alcohol) (average Mw 9,000–10,000, 80 % hydrolyzed).
   (b) PCL (average Mw 65,000, average Mn 42,500, pellets).
   (c) Sodium chloride (crystalline, fine).
   (d) Ethyl acetate.

(e) Deionized water.

(f) Glass scintillation vials with caps.

(g) Mechanical sieve tower with 250, 150, 75, and 38 μm pore sizes (Fisher Scientific—Pittsburgh, PA, USA).

(h) Glass capillary mandrels, 1.5 mm diameter (World Precision Instruments—Sarasota, Fl, USA).

(i) Vortex Genie 2 (USA Scientific—Ocala, FL, USA).

(j) Excess cardboard.

2.2 Animals and Surgical Equipment

1. Lewis rat: Albino females weighing >250 g (Harlan Laboratories—Indianapolis, IN, USA) are used for sciatic nerve transection and repair experiments (*see* **Note 1**). The animals are maintained under standard animal care conditions (12:12-h light/dark cycle, lights coming on at 7 a.m., room temperature 21 °C) with food and water ad libitum. Prior to the start of any experiment, the health of the rodents is to be assessed according to published guidelines [16]. All protocols require approval under the appropriate governing Institutional Review Board (IRB) and Institutional Animal Care and Use Committee (IACUC) regulatory panels as established by the 1986 US Animal Welfare Act.
2. Anesthetic solution: Stock solutions of Ketathesia (ketamine) 100 mg/ml (Butler Schein—Dublin, OH, USA) and Anased (xylazine) 20 mg/ml (Lloyd Laboratories—Shenandoah, IA, USA) are used to prepare the anesthetic solution. Target doses per animal for ketamine and xylazine are 80 mg/kg and 10 mg/kg, respectively (*see* **Note 2**).
3. Postoperative analgesia: Animals are treated postoperatively for 72 h. Initial dose is given at the time of surgery and repeated every 12 h thereafter. Buprenex (buprenorphine) 0.3 mg/ml stock solution in 1 ml ampule vials (Reckitt Benckiser Pharmaceuticals—Richmond, VA, USA) is used to relieve and/or prevent signs and symptoms of pain and/or distress for postoperatively. Target dose per animal of buprenorphine is 0.1 mg/kg (*see* **Note 3**).
4. An operating microscope (Carl Zeiss Meditec—Dublin, CA, USA) or surgical loupes (Surgical Acuity—Middleton, WI, USA) are required for microsurgical implantation of nerve conduit.
5. Surgical instruments: #10 scalpel (Aspen Surgical—Caledonia, MI, USA), Brown–Adson forceps, iris scissors, self-retaining retractors, needle driver, suture scissors (ASSI—Westbury, NY, USA).
6. Microsurgical instruments: Straight scissors, needle driver, curved or straight microsurgical forceps (ASSI—Westbury, NY, USA).

7. Suture: 9-0 nylon (Microsurgery Instruments—Bellaire, TX, USA) used for microsurgical implantation of the nerve guide and 4-0 Vicryl (Ethicon—Somerville, NJ, USA) used for layered muscle and skin closures.
8. 70 % isopropyl alcohol wipes (Becton Dickinson—Franklin Lakes, NJ, USA).
9. Veterinary hair clippers (Wahl Clipper Corporation—Sterling, IL, USA).
10. Electrocautery forceps (Covidien—Boulder, CO, USA).

3 Methods

3.1 PCL Nerve Guide Fabrication

1. Day 1: Coat glass capillary mandrels with a 17 % w/v solution of aqueous poly(vinyl alcohol) and allow to dry overnight (*see* **Note 4**).
2. Day 2: Make a 7 % w/v solution of PCL in ethyl acetate. Stir solution in heated oil bath (40 °C). PCL/ethyl acetate mixture should take a couple of hours to dissolve.
3. Porosity percentage is determined by a volume ratio of NaCl to PCL. Create an 80 % v/v solution of NaCl:PCL in ethyl acetate (*see* **Note 5**).
4. Immerse PVA-coated glass capillaries into the NaCl/PCL slurry (*see* **Note 6**).
5. Allow 10 min for ethyl acetate to evaporate between each successive dipping. Glass capillary mandrels should be dipped six times for a wall thickness of 600 μm.
6. Submerge mandrels with polymer coating in distilled, deionized water overnight to allow for salt and PVA dissolution. Previous work in our laboratory suggests that a minimum salt dissolution time of 5 h can also be utilized [13].
7. After a minimum 5-h salt leaching time, gently slide PCL tubes off of the glass capillary mandrels. Cut into 1.7 cm in length.
8. Nerve guides can be sterilized with a room temperature ethylene oxide machine. This method does not appear to alter porosity or mechanical strength of the guide (unpublished results—*see* Fig. 1).

3.2 Sciatic Nerve Injury and Repair

1. Administer the rat an intraperitoneal injection of ketamine (100 mg/ml) and xylazine (20 mg/ml) and return to housing cage until fully anesthetized (*see* **Note 7**).
2. Shave right lower extremity with veterinary clippers, and cleanse the area with an alcohol prep pad.
3. Position the rat on the operating table with left side down and head facing away from operating surgeon (*see* **Note 8**).

Fig. 1 Scanning electron microscope image of PCL nerve guide with embedded microspheres. A number of other growth factor delivery systems can be used with this PCL guide

4. Make 3 cm incision parallel to and 5 mm inferior to the femur on the lateral aspect of the right leg with a #10 scalpel or scissors (*see* **Note 9**).
5. Elevate the skin and underlying panniculus carnosus muscle off the underlying leg musculature by inserting the points of the iris scissors and spreading parallel to the undersurface of the skin.
6. Locate the biceps femoris muscle, and spread its fibers apart perpendicular to the long axis of the muscle belly, creating a small window through the muscle. The sciatic nerve will be encountered just deep to the muscle in a clean tissue plane (*see* **Note 10**).
7. Open the muscle with a combination of spreading and cutting both proximally and distally to completely expose the sciatic nerve deep to the biceps femoris muscle. Insert self-retaining retractors both proximally and distally to aid in the process and to secure the final exposure (*see* Fig. 2).
8. Bring in surgical microscope or begin using surgical loupe magnification (*see* **Note 11**).
9. Using microsurgical forceps and scissors, circumferentially clean the adherent tissue from the nerve by carefully spreading parallel to the nerve fibers, freeing up the segment from its exit out of the pelvis to just proximal to its distal bifurcation.

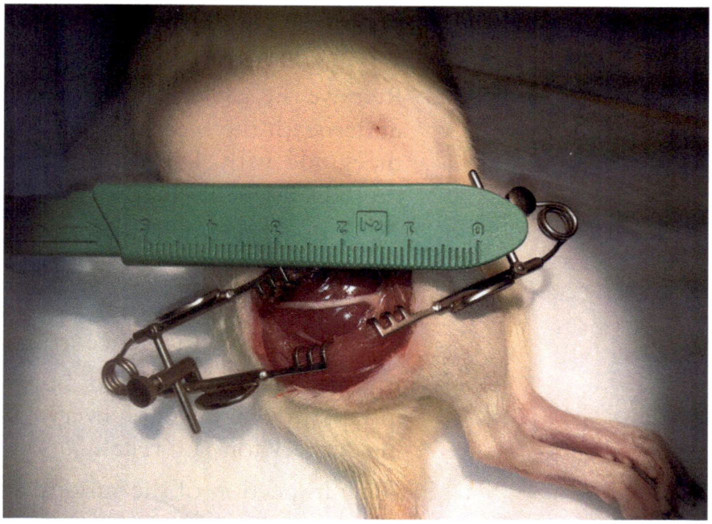

Fig. 2 Surgical exposure of the sciatic nerve in the rat right hind limb. Note the placement of the self-retaining retractors that secure exposure for the entire length of the nerve

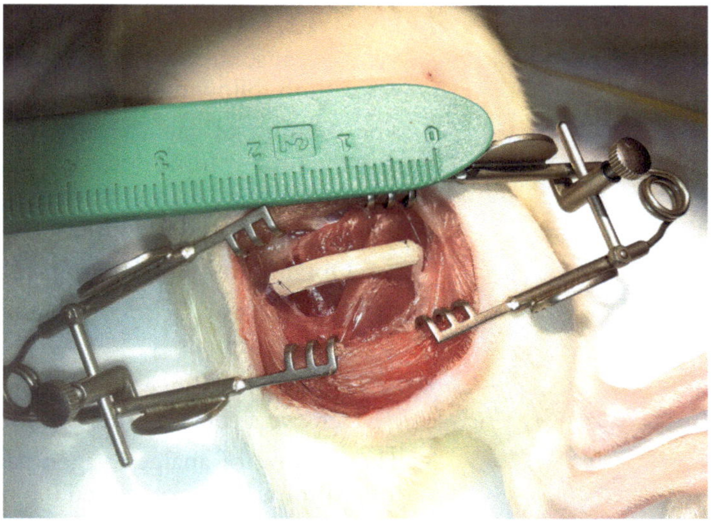

Fig. 3 Successfully implanted nerve guide in the rat right hind limb. Note the "U" stitches securing the nerve stumps 1 mm from the edge of the guide

10. Mark and trim a 5 mm segment from the middle portion of the nerve, being careful to leave at least 3 mm of nerve proximal to the distal bifurcation. The severed nerve stumps will retract slightly (*see* **Note 12**).
11. Secure the proximal portion of the nerve conduit with two "U" stitches oriented 180° from one another (*see* Fig. 3).

12. Secure the distal portion of the nerve conduit with two "U" stitches oriented 180° away from one another.
13. Remove the self-retaining retractors, and tuck the nerve guide underneath the biceps femoris muscle. Close the muscle over the guide with 4-0 vicryl suture using three or four interrupted sutures to reapproximate the split in the muscle.
14. Close the skin with a single running 4-0 vicryl suture.
15. Administer intraperitoneal injection of buprenorphine (0.1 mg/kg) and return the animal to its cage. Monitor closely to ensure a safe emergence from anesthesia (*see* **Note 13**).
16. Animals are administered buprenorphine (0.1 mg/kg) via intraperitoneal injection every 12 h for a total of 72 h following surgery for pain relief.
17. Daily inspection of the surgical site for signs of wound dehiscence, infection, or extrusion of the nerve guide is recommended until full healing has occurred.

4 Notes

1. Lewis rats, in particular females, should be used for the experiment because they have less of tendency to chew on the denervated limb following surgery than other strains.
2. The dilution of each anesthetic agent should be done such that the total volume does not exceed 0.5 ml per injection. For example, a standard solution used to anesthetize ten 250 g rats would be prepared as follows:
   (a) Obtain 2.0 ml of 100 mg/ml ketamine solution.
   (b) And add the above solution to 1.25 ml of 20 mg/ml xylazine solution.
   (c) Dilute the above mixture with 1.75 ml sterile PBS to achieve a total volume of 5.0 ml.
3. The dilution of analgesic agent should be done such that the total volume does not exceed 0.5 ml per injection. For example, a standard solution used to treat ten 250 g rats for a 72-h period (i.e., six doses for each rat) would be prepared as follows:
   (a) Obtain 4.98 ml of 0.3 mg/ml buprenorphine solution.
   (b) And add the above to 25.02 ml of sterile PBS to achieve a total volume of 30.0 ml.
4. To prepare for PCL nerve guide fabrication, glass capillary mandrels should be glued to flat pieces of cardboard with a glue gun. Glass capillary mandrels should be completely perpendicular to the cardboard with no tilting. This creates uniform wall thickness along the entire length of the guide as the PCL/NaCl coating dries between successive dips.

5. Previous work in our laboratory has established that a wall thickness of 600 μm with an 80 % porosity percentage and a pore size of 10–38 μm are optimal in retaining trophic factors within the conduit while also not limiting glucose diffusion into the guide [13]. Salt should be finely sifted using a mechanical sieve tower consisting of a 250, 150, 75, and 38 μm sieves. The portion of salt that goes through the 38 μm sieve should be saved and used for nerve guide fabrication. However, pore size, wall thickness, and porosity percentage should be optimized with diffusion studies if different luminal fillers within the guide are utilized.
6. NaCl/PCL slurry should be vortexed vigorously before dipping to ensure homogenous salt distribution throughout the PCL mixture. Asymmetric pore distribution in the final nerve guide product can result if polymer slurry is not vortexed effectively.
7. When administering an intraperitoneal injection, it is advantageous to firmly grasp the rat with the head between the index and long finger with the animal's back secured against the palm of the surgeon's hand. The tail of the rat can be wrapped behind the thumb and secured with the index finger. The rat can then be picked up with the belly fully exposed to the surgeon, limiting the chance for an accidental needle stick or an inaccurate delivery of medications.
8. Always perform the sciatic nerve transection on the right leg to avoid confusion later during analysis.
9. Rather than incising with a scalpel, make a small cut on the proximal portion of the leg and spread with scissors underneath the entire length of the proposed incision. Then open the scissors and use the subcutaneous tunnel to cut the entire length of the incision. This will save time and provide a cleaner edge for closure.
10. There is a small vessel that will always bleed when splitting the biceps femoris muscle proximally in the rat hind limb. Be certain to have bipolar electrocautery at the ready to extinguish any encountered bleeding.
11. If using surgical loupes, they can be used for the entire procedure. If a surgical microscope is used, it is best used for only the suturing of the nerve guide.
12. Sometimes the sciatic nerve will bifurcate relatively high in its course behind the biceps femoris muscle. If this is the case, take a 5 mm segment closer to the exit of the nerve from the pelvis.
13. If the animals are taking excessive amounts of time to emerge from anesthesia, the dosages of ketamine and xylazine can be halved. However, this will limit the amount of time for the implantation and should only be done by those having significant experience with the procedure.

References

1. Siemionow M, Bozkurt M, Zor F (2010) Regeneration and repair of peripheral nerves with different biomaterials: review. Microsurgery 30(7):574–588. doi:10.1002/micr.20799
2. Lundborg G (2000) A 25-year perspective of peripheral nerve surgery: evolving neuroscientific concepts and clinical significance. J Hand Surg [Am] 25(3):391–414. doi:10.1053/jhsu.2000.4165
3. Ngeow WC (2010) Scar less: a review of methods of scar reduction at sites of peripheral nerve repair. Oral Surg Oral Med Oral Pathol Oral Radiol Endod 109(3):357–366, Doi:S1079-2104(09)00429-6 [pii] 10.1016/j.tripleo.2009.06.030
4. Dow DE, Cederna PS, Hassett CA, Kostrominova TY, Faulkner JA, Dennis RG (2004) Number of contractions to maintain mass and force of a denervated rat muscle. Muscle Nerve 30(1):77–86
5. Battal MN, Hata Y (1997) A review on the history of end-to-side neurorrhaphy. Plast Reconstr Surg 99(7):2110–2111
6. Ichihara S, Inada Y, Nakamura T (2008) Artificial nerve tubes and their application for repair of peripheral nerve injury: an update of current concepts. Injury 39(Suppl 4):29–39, Doi:S0020-1383(08)00399-9 [pii] 10.1016/j.injury.2008.08.029
7. Myckatyn T, MacKinnon S (2004) A review of research endeavors to optimize peripheral nerve reconstruction. Neurol Res 26(2):124–138
8. Nichols CM, Brenner MJ, Fox IK, Tung TH, Hunter DA, Rickman SR, Mackinnon SE (2004) Effects of motor versus sensory nerve grafts on peripheral nerve regeneration. Exp Neurol 190(2):347–355. doi:10.1016/j.expneurol.2004.08.003
9. Willerth SM, Sakiyama-Elbert SE (2007) Approaches to neural tissue engineering using scaffolds for drug delivery. Adv Drug Deliv Rev 59(4–5):325–338
10. Ghaznavi AM, Kokai LE, Lovett ML, Kaplan DL, Marra KG (2011) Silk fibroin conduits: a cellular and functional assessment of peripheral nerve repair. Ann Plast Surg 66(3):273–279. doi:10.1097/SAP.0b013e3181e6cff7
11. Pfister LA, Papaloïzos M, Merkle HP, Gander B (2007) Nerve conduits and growth factor delivery in peripheral nerve repair. J Peripher Nerv Syst 12(2):65–82
12. Rovak JM, Mungara AK, Aydin MA, Cederna PS (2004) Effects of vascular endothelial growth factor on nerve regeneration in acellular nerve grafts. J Reconstr Microsurg 20(1):53–58. doi:10.1055/s-2004-818050
13. Kokai LE, Lin Y-C, Oyster NM, Marra KG (2009) Diffusion of soluble factors through degradable polymer nerve guides: controlling manufacturing parameters. Acta Biomater 5(7):2540–2550
14. Nectow AR, Marra KG, Kaplan DL (2012) Biomaterials for the development of peripheral nerve guidance conduits. Tissue Engineering Part B, Reviews 18(1):40–50
15. Lin YC, Ramadan M, Hronik-Tupaj M, Kaplan DL, Philips BJ, Sivak W, Rubin JP, Marra KG (2011) Spatially controlled delivery of neurotrophic factors in silk fibroin-based nerve conduits for peripheral nerve repair. Ann Plast Surg 67(2):147–155. doi:10.1097/SAP.0b013e3182240346
16. Morton DB, Griffiths PH (1985) Guidelines on the recognition of pain, distress and discomfort in experimental animals and an hypothesis for assessment. Vet Rec 116(16):431–436

Chapter 12

A Highly Reproducible Mouse Model of Compression Spinal Cord Injury

Suelen Adriani Marques, Fernanda Martins de Almeida, Klauss Mostacada, and Ana Maria Blanco Martinez

Abstract

Experimental spinal cord injury (SCI) can maintain the continuity of the spinal cord, as in the contusion (e.g., weight-fall) or compression models, or not, when there is a partial or a complete transection. The majority of acute human SCI is not followed by complete transection, but there is a combination of contusion, compression, and possibly partial transection. The method described here is a compressive mouse model that presents a combination of contusion and compression components and has many facilities in its execution. This lesion was established by our group and represents a simple, reliable, and inexpensive clip compression model with functional and morphological reproducibility. In this chapter we describe, step by step, the protocol of this experimental SCI.

Key words Spinal cord injury, Vascular clip, Compressive injury, Laminectomy, Mouse

1 Introduction

Spinal cord injury (SCI) is considered a serious neurological disorder that results from a trauma affecting mainly young adults, leading to disabilities that remain throughout the rest of their lives. The lesion causes sensory, autonomous, and motor deficits due to neuronal and glial cell death, axonal damage, and demyelination, and these events are responsible for functional loss beyond the injury site. According to the National Spinal Cord Injury database (2011), the incidence of SCI in the USA is approximately 40 cases per million population with 12,000 new cases every year [1].

Despite the advances in surgical protocols, general care, and rehabilitation techniques, efficient therapies that can be successfully translated into clinical setting are still not available [2–4]. In the last two decades, several groups investigated regenerative strategies in central nervous system [5–7] and showed promising results for attenuating the devastating effects after SCI. However one

Andrew J. Murray (ed.), *Axon Growth and Regeneration: Methods and Protocols*, Methods in Molecular Biology, vol. 1162, DOI 10.1007/978-1-4939-0777-9_12, © Springer Science+Business Media New York 2014

point that makes the comparison among all the studies difficult is the diversity of the type of lesion used in the experiments; another problem is that some methods require expensive instruments and a skilled person to handle it.

There is a variety of type of lesions that have been used as a model of SCI such as balloon inflation [8, 9], compression [10–14], crush with forceps [15], lateral hemisection [16, 17], full transection [18], dorsal column transection [19, 20], contusion [21–23], and removal of spinal cord segment [24–26]. Despite this great variety, the model selected must be appropriate to answer the questions raised by the researchers and should be easily reproduced by other groups. For this reason, our group established an easy, cheap, and highly reproducible model of compressive SCI in mice and used it to test and compare several types of treatments [12–14].

For that lesion we use a vascular clip, with a known pressure force, that provides a compression in the spinal cord, keeping the dura mater intact. The extra dural compression using a clip was introduced by Rivlin and Tator [27] and simulates the majority of lesion in humans, because it is composed by both contusion and compression. This lesion model presents some advantages such as high reproducibility, low cost, and no need for previous training to use it.

2 Materials

2.1 Animals

Female C57 Bl/6 mice around 10 weeks old (weighting 20 g) can be used for this surgery. The animals must be maintained under standard animal care conditions (12:12-h light/dark cycle, lights coming on at 7 a.m., room temperature 21 °C), with food and water ad libitum, before and after surgery. Prior to the start of any experiment, the locomotor function must be assessed to detect changes and select the healthy mice. All experimental procedures need to be approved by the Local Ethical Commission of Animal Care.

2.2 Neurosurgery Materials

Alcohol 70 %, sterile cotton wool and sterile mounted swab, magnetic fixator retraction system (Fine Science Tools, 18200-20), scalpel and scalpel blades, retractor, anatomical and surgical tweezers, microrongeur (0.5 × 12 mm, Fine Science Tools, 16221-14), 1 mL disposable syringes with 26-G needle for intraperitoneal (i.p.) injection of the anesthetic solution, micro scissors, two surgical forceps (5C, Dumont), vessel clip (with 30 × *g* force, Kent Scientific Corporation, INS 14120, USA), 6/0 monofilament nylon with suture needles and needle holder, 3 mL disposable syringes with 26-G needle for subcutaneous (s.c.) saline injection plus prophylactic Baytril (2.5 mg/kg/d, s.c., Bayer, Brazil; [12]).

3 Methods

This extra dural clip compression injury model uses a clip that is closed around the spinal cord in the vertical plane and therefore exerts force from side to side on the spinal cord. Clip closing in the vertical plane was also used by von Euler and co-workers [28], and the results are compatible with those described in the literature for clips that close in the anteroposterior plane ([27, 29, 30]; *see* **Note 1**). In this model, the spinal cord damage simulates the clinical situation seen in the majority of human cases and also presents a low cost.

All the steps of surgical procedures are illustrated in Figs. 1 and 2.

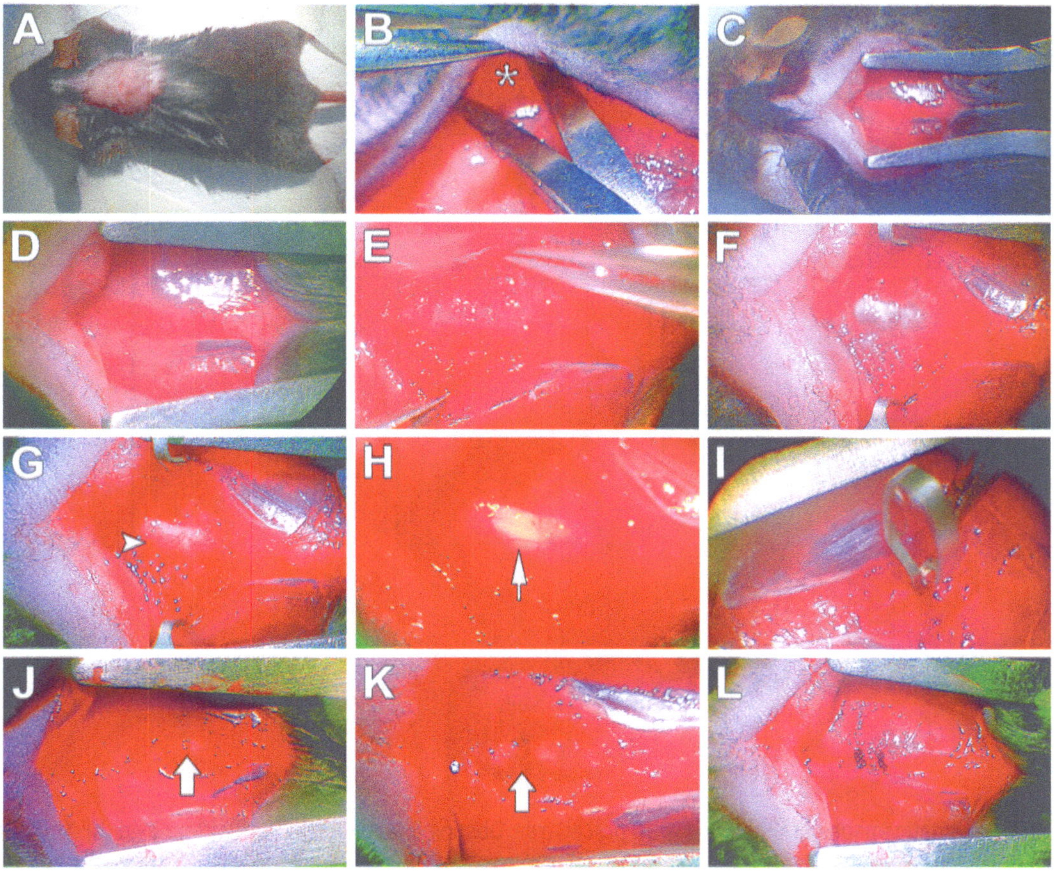

Fig. 1 Images showing the main steps of SCI surgical procedures. In (**a**) we can observe the animal anesthetized. After the incision on the skin along the back midline (**a**), the subcutaneous tissue is divulsioned with a micro scissor (**b**, *asterisk*) and the retractor is inserted (**c**, **d**). Paravertebral muscles are incised (**e**) until the vertebrae exposition (**f**). The spine process is removed with a microrongeur, and the laminas are dissected out using a forceps (**g**, *arrowheads*). Observe that the spinal cord is exposed (**h**, *arrows*), followed by clip compression for 1 min (**i**). (**j**) and (**k**) represent the hemorrhagic ring (*solid arrow*) in different magnification after compression. After lesion, the muscle (**l**) and skin are sutured

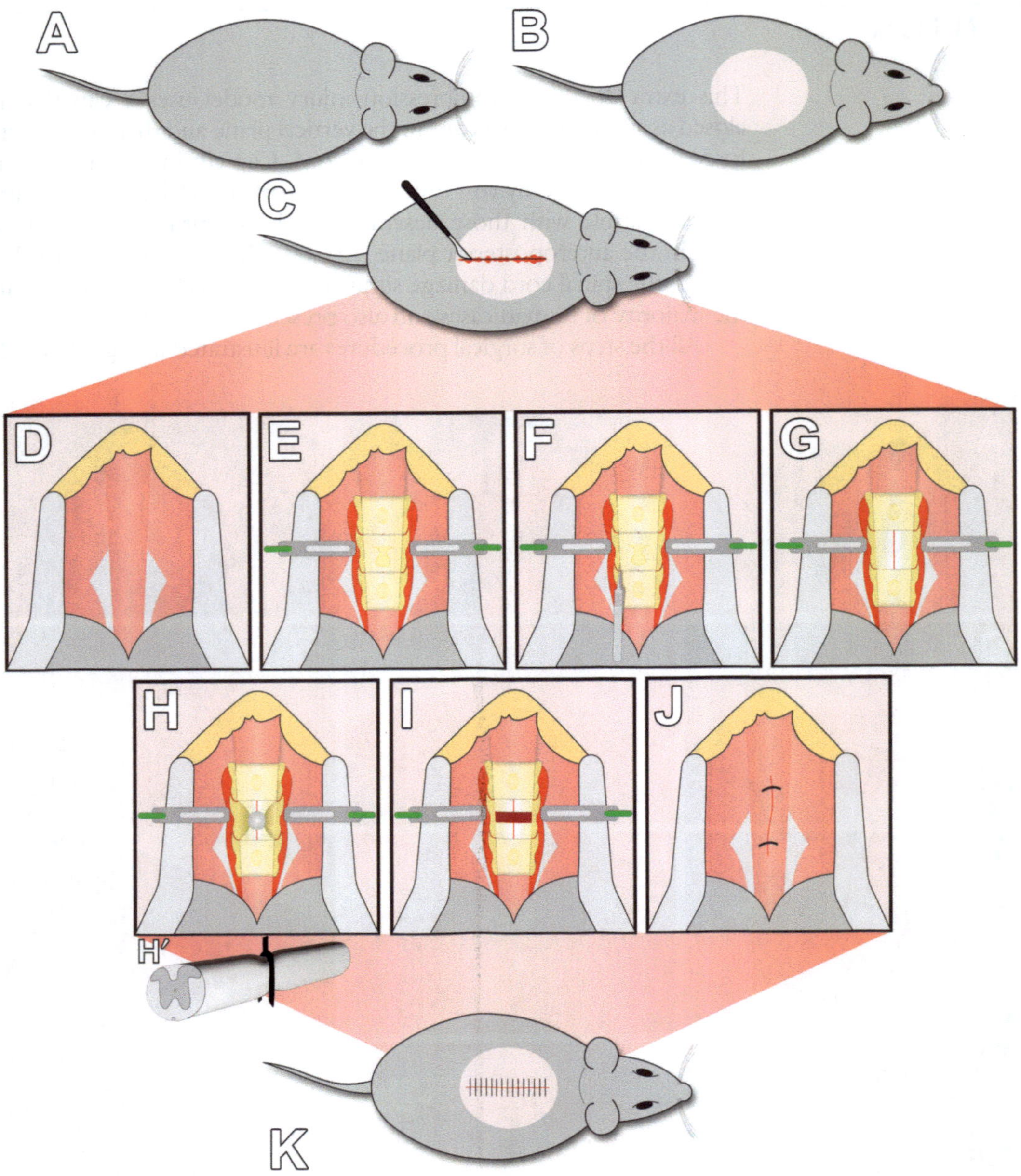

Fig. 2 Schematic images of the SCI surgical procedures. Observe the mice (**a**) after hair removal (**b**). The incision is made along the midline with a scalpel (**c**), the retractor is inserted (**d**) to expose surgical field, and paravertebral muscles are removed between T8 and T10 (**e**). Laminae are dissected out using a forceps like described in **step 5** (**f**), and after that the spinal cord is exposed (**g**). A superior view of clip compression around spinal cord (**h**, *see* **step 6** and **Note 1**), and H′ illustrates the side-to-side compression exerted by the clip. After clip removal the hemorrhagic ring is evident (**i**). Images (**j**) and (**k**) show muscle and skin sutures in two layers (*see* **step 9**)

1. Animals are deeply anesthetized, placed in a hot pad, and immobilized to start the surgery.
2. The skin is incised along the midline of the back with a scalpel, and the subcutaneous tissue is divulsioned from the muscle tissue with the aid of a micro scissors. At this point the retractor is inserted to facilitate the surgical field view.
3. Superficial and deep paravertebral muscles are incised parallel to the spine processes with a scalpel until the vertebrae are exposed at the thoracic level (T8–T10; *see* **Note 2**).
4. Ligaments (flavum, interspinous, intertransverse, and supraspinous) and the small muscles (such as multifidus, intertransversarii, and interspinales muscles) must be removed between the T8 and T10 vertebrae before the laminectomy.
5. Laminectomy is performed just at the T9 level. Under visual guidance, use an operating microscope (F104, OPTO, Brazil). The T9 spine process is dissected out using the microrongeur to remove the spine process, and then the bilateral laminas are dissected out with the forceps help. Be careful in this step to avoid damage to the dura mater (*see* **Note 3**). The forceps should be introduced in the vertebrae channel with the tip to the top, next to the laminae, forming a 45° angle.
6. Bilateral transverse processes need to be dissected out to allow the vessel clip entry into the vertebral channel.
7. 1-min extradural compression with vessel clip (30 × *g* pressure, Kent Scientific Corporation, INS 14120, USA) should be performed around the exposed spinal cord in order to cause a moderately severe acute-compression injury (*see* **Notes 4** and **5**).
8. After removal of the clip, the epicenter of the lesion has an evident hemorrhagic ring at the site of the compression with preservation of the dura mater.
9. To finish, muscles and skin are sutured in two different layers (deep layer muscle, and superficial layer subcutaneous tissue plus skin) using 6/0 monofilament nylon with micro needle holder. By doing this there is less adherence, facilitating a new surgical intervention, if needed.
10. Immediately after surgery, an antibacterial spray should be applied topically on the surgical scar and the animals should receive a subcutaneous saline injection (1–2 mL) and prophylactic antibiotic (Baytril, 2.5 mg/kg/d, s.c., Bayer, Brazil) to replace the blood loss and prevent infection.
11. The animals should be left to recover on a warm pad until thermoregulation and an alert state is reestablished. Then they can be returned to their home cages, with free access to food and water.
12. Operated animals should receive manual bladder expression twice a day until the return of bladder function; and they also should receive appropriate veterinary care when needed.

4 Notes

1. The vascular clip is closed around the spinal cord in the vertical plane and therefore exerts force from side to side on the spinal cord. Despite the form that the clip is closed around the spinal cord, functional and morphological results are compatible with those described in the literature for clips that close in the anteroposterior plane [12, 28] as first introduced by Rivlin and Tator [27].
2. Mice have 13 thoracic vertebrae that increase in size as one proceeds down the spine. T9 vertebra has certain particularities, its laminae are large and overlapping with the T10 vertebra, and its spinous process is projected over the lower vertebra. Also, T9 has a single superior costal demi-facet above which provides articular spaces for ribs. We used the inferior fibers of the trapezius as the reference to make the surgical incision. T9 vertebra was found observing these anatomical details and by palpation of the spinous process to detect the difference in the space between T9–T10 and T10–T11 vertebrae (Fig. 3).
3. Be careful during the laminectomy procedure because the dura mater is fragile and must be maintained without damage to avoid bias in the study. Furthermore, one should be careful when introducing the clip into the vertebral channel, so that it goes as far as until the tip of the blades touches the anterior

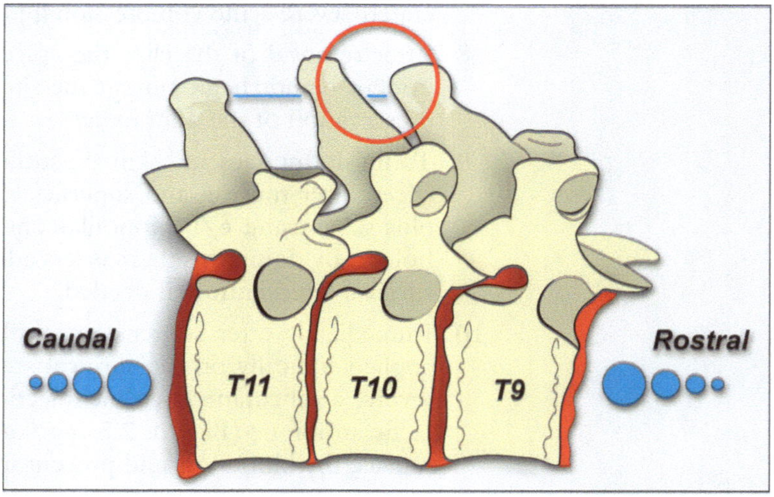

Fig. 3 Schematic representation of the vertebrae T9–T11. The *blue line* highlights the difference in space between the spinous processes of T9–T11 that are observed during palpation of these. The *red circle* shows how close to each other these spinous processes are

body of the vertebra. This care can prevent sliding of the clip and avoids a wrong compression, just promoting a dorsal spinal cord compression.

4. The blades of the vascular clip used in this model can suffer deformation after being used, causing some alterations especially in the amount of force exerted on the spinal cord. The clip must be assessed before its use; if the blades are deformed it should be discarded. In average a clip can be used five times without deformation, but you should always check it before using it.
5. Changes in compression time can cause spinal cord injury with different impairment degree, less or more severe, and therefore changing the functional and morphological results. In addition, the hemorrhagic ring aspect that appears immediately after the withdrawal of the clip will be different. Also, variation on the clip compressive force could modulate the severity of the SCI.

References

1. Statistical Center (NSCIS). Spinal cord injury facts and figures at a glance. https://www.nscisc.uab.edu/PublicDocuments/nscisc_home/pdf/Facts%202011%20Feb%20Final.pdf. Accessed Feb 2011
2. Andres RH, Meyer M, Ducray AD et al (2008) Restorative neuroscience: concepts and perspectives. Swiss Med Wkly 138:155–172
3. Eftekhar E, Karimi-Abdolrezaee S, Fehlings MG (2008) Current status of experimental cell replacement approaches to spinal cord injury. Neurosurg Focus 24:E18
4. Vawda R, Fehlings MG (2013) Mesenchymal cells in the treatment of spinal cord injury: current & future perspectives. Curr Stem Cell Res Ther 8:25–38
5. Cizkova D, Novotna I, Slovinska L et al (2011) Repetitive intrathecal catheter delivery of bone marrow mesenchymal stromal cells improves functional recovery in a rat model of contusive spinal cord injury. J Neurotrauma 28(9):1951–1961
6. Tetzlaff W, Okon EB, Karimi-Abdolrezaee S et al (2011) A systematic review of cellular transplantation therapies for spinal cord injury. J Neurotrauma 28(8):1611–1682
7. Zhou Z, Chen Y, Zhang H et al (2013) Comparison of mesenchymal stromal cells from human bone marrow and adipose tissue for the treatment of spinal cord injury. Cytotherapy 15(4):434–482
8. Martin D, Robe P, Franzen R et al (1996) Effects of Schwann cell transplantation in a contusion model of rat spinal cord injury. J Neurosci Res 45:588–597
9. Cizkova D, Rosocha J, Vanicky I et al (2006) Transplants of human mesenchymal stem cells improve functional recovery after spinal cord injury in the rat. Cell Mol Neurobiol 26:1167–1180
10. Boyd JG, Lee J, Skihar V et al (2004) LacZ-expressing olfactory ensheathing cells do not associate with myelinated axons after implantation into the compressed spinal cord. Proc Natl Acad Sci U S A 101:2162–2166
11. Firouzi M, Moshayedi P, Saberi H et al (2006) Transplantation of Schwann cells to subarachnoid space induces repair in contused rat spinal cord. Neurosci Lett 402:66–70
12. Marques SA, Garcez VF, Del Bel EA et al (2009) A simple, inexpensive and easily reproducible model of spinal cord injury in mice: Morphological and functional assessment. J Neurosci Methods 177:183–193
13. Marques SA, Almeida FM, Fernandes A et al (2010) Predifferentiated embryonic stem cells promote functional recovery after spinal cord compressive injury. Brain Res 1349:115–128
14. Almeida FM, Marques SA, Ramalho BS et al (2011) Human dental pulp cells: a new source of cell therapy in a mouse model of compressive spinal cord injury. J Neurotrauma 28:1939–1949
15. Papastefanaki F, Chen P, Lavdas AA et al (2007) Grafts of Schwann cells engineered to

express PSA-NCAM promote functional recovery after spinal cord injury. Brain 130: 2159–2174

16. Xu XM, Zhang SX, Li H et al (1999) Regrowth of axons into the distal spinal cord through a Schwann-cell-seeded mini-channel implanted into hemisected adult rat spinal cord. Eur J Neurosci 11:1723–1740
17. Hsu JY, Xu XM (2005) Early profiles of axonal growth and astroglial response after spinal cord hemisection and implantation of Schwann cell-seeded guidance channels in adult rats. J Neurosci Res 82:472–483
18. Guest JD, Herrera L, Margitich I et al (2008) Xenografts of expanded primate olfactory ensheathing glia support transient behavioral recovery that is independent of serotonergic or corticospinal axonal regeneration in nude rats following spinal cord transection. Exp Neurol 212:261–274
19. Imaizumi T, Lankford KL, Kocsis JD (2000) Transplantation of olfactory ensheathing cells or Schwann cells restores rapid and secure conduction across the transected spinal cord. Brain Res 854:70–78
20. Vroemen M, Caioni M, Bogdahn U et al (2007) Failure of Schwann cells as supporting cells for adult neural progenitor cell grafts in the acutely injured spinal cord. Cell Tissue Res 32:1–13
21. Hill CE, Hurtado A, Blits B et al (2007) Early necrosis and apoptosis of Schwann cells transplanted into the injured rat spinal cord. Eur J Neurosci 26:1433–1445
22. Pearse DD, Sanchez AR, Pereira FC et al (2007) Transplantation of Schwann cells and = or olfactory ensheathing glia into the contused spinal cord: survival, migration, axon association, and functional recovery. Glia 55: 976–1000
23. Schaal SM, Kitay BM, Cho KS et al (2007) Schwann cell transplantation improves reticulospinal axon growth and forelimb strength after severe cervical spinal cord contusion. Cell Transplant 16:207–228
24. Pinzon A, Calancie B, Oudega M et al (2001) Conduction of impulses by axons regenerated in a Schwann cell graft in the transected adult rat thoracic spinal cord. J Neurosci Res 64: 533–541
25. Plant GW, Bates ML, Bunge MB (2001) Inhibitory proteoglycan immunoreactivity is higher at the caudal than the rostral Schwann cell graft-transected spinal cord interface. Mol Cell Neurosci 17:471–487
26. Vavrek R, Pearse DD, Fouad K (2007) Neuronal populations capable of regeneration following a combined treatment in rats with spinal cord transection. J Neurotrauma 24: 1667–1673
27. Rivlin AS, Tator CH (1978) Effects of duration of acute spinal cord compression in a new acute cord injury model in the rat. Surg Neurol 10(1):38–43
28. von Euler M, Seiger A, Sundstrom E (1997) Clip compression injury in the spinal cord: a correlative study of neurological and morphological alterations. Exp Neurol 145: 502–510
29. Khan M, Griebel R (1983) Acute spinal cord injury in the rat: comparison of three experimental techniques. Can J Neurol Sci 10: 161–165
30. Poon PC, Gupta D, Shoichet MS, Tator CH (2007) Clip compression model is useful for thoracic spinal cord injuries. Spine 32(25): 2853–2859

Chapter 13

Using Templated Agarose Scaffolds to Promote Axon Regeneration Through Sites of Spinal Cord Injury

Jacob Koffler, Ramsey F. Samara, and Ephron S. Rosenzweig

Abstract

The past 30 years of research in spinal cord injury (SCI) have revealed that, under certain conditions, some types of axons are able to regenerate. To aid these axons in bridging the lesion site, many experimenters place cellular grafts at the lesion. However, to increase the potential for functional recovery, it is likely advantageous to maximize the number of axons that reach the intact spinal cord on the other side of the lesion. Implanting linear-channeled scaffolds at the lesion site provides growing axons with linear growth paths, which minimizes the distance they must travel to reach healthy tissue. Moreover, the linear channels help the regenerating axons maintain the correct mediolateral and dorsoventral position in the spinal cord, which may also improve functional recovery by keeping the axons nearer to their correct targets. Here, we provide a protocol to perform a full spinal cord transection in rats that accommodates an implanted scaffold.

Key words Spinal cord injury, Scaffold, Regeneration, Axons, Surgery, Complete transection

1 Introduction

There are no effective therapies for promoting regeneration after acute or chronic spinal cord injury (SCI) in humans. However, research in animal models has revealed that, under the right conditions, axons in the spinal cord may regenerate. One of many experimental approaches to promote regeneration in animal models is to graft cells such as astrocytes, Schwann cells, marrow stromal cells, or stem cells at injury sites [1–6]. However, these cell grafts do not allow for organized, directed growth of injured axons through an injury site, but rather result in random axonal trajectories. Random growth results in a longer path (and therefore delay) to reach uninjured tissue and an increased risk of mistargeting, both of which may hinder functional recovery. Indeed, most axons do not regenerate through the lesion site into the host tissue beyond, and functional recovery after these treatments is limited [1, 3–6]. Thus, guiding regenerating axons

Andrew J. Murray (ed.), *Axon Growth and Regeneration: Methods and Protocols*, Methods in Molecular Biology, vol. 1162,
DOI 10.1007/978-1-4939-0777-9_13, © Springer Science+Business Media New York 2014

across the lesion site to their appropriate targets may greatly improve functional recovery after SCI.

In the last 5 years, we have succeeded in orienting and guiding in vivo axonal regeneration using biocompatible synthetic scaffolds that do not cause significant inflammation [2, 7, 8]. These scaffolds are engineered with linear channels and create a direct bridge between the rostral and the caudal aspects of the lesion when implanted [2]. Remarkably, this guidance can be achieved across a complete spinal cord transection site, perhaps the most challenging model of SCI. This model creates a clean gap in the spinal cord using a combined cut-and-aspirate technique. Briefly, the rostral and caudal aspects of the lesion are created by sharp, clean cuts that result in smooth lesion boundaries. The spinal cord between these smooth cuts is removed via aspiration with minimal manipulation of the rostral and caudal lesion boundaries. This results in a severe lesion that injures ascending and descending tracts and drastically impairs motor function (as measured by tests such as the Basso, Beattie, and Bresnahan open field locomotion test) [9]. Moreover, this lesion allows the implanted scaffold to make good contact with uninjured host tissue. Finally, the full transection model allows unequivocal identification of axon regeneration across a lesion site. This stands in contrast to contusive models, in which axon growth can originate from cut axons (regeneration) or intact axons (sprouting). The methods for producing this lesion and implanting a scaffold are described herein.

2 Materials

2.1 Scaffolds

We use 3 % agarose hydrogel scaffolds containing linear channels, manufactured as described earlier [2, 7, 8]. Scaffolds should be rinsed for 20 min prior to surgery in sterile PBS or saline. Other types of scaffold may require different handling methods.

2.2 Anesthetic Cocktail

Ketamine: 40 mg/ml, Dexdomitor: 0.3 mg/ml. For every 10 ml of anesthetic, use 4 ml Ketamine (100 mg/ml) + 6 ml Dexdomitor (0.5 mg/ml).

2.3 Reversing Agent

Antisedan: 1.25 mg/ml. Replace 2.5 ml of saline in a 10 ml vial with 5 mg/ml Antisedan.

2.4 Postsurgical Drugs and Solutions

Banamine (analgesic): 0.5 mg/ml. Replace 0.5 ml from a 50 ml saline vial with 50 mg/ml Banamine.

Ampicillin (antibiotic): 40 mg/ml. Use 10 ml from a 50 ml saline vial to hydrate two vials of lyophilized Ampicillin (1 g each). Return the hydrated solution to the saline vial.

Lactated Ringer's Solution (fluid replacement): 3–5 ml SC.

2.5 Surgery Tools

Friedman-Pearson Rongeur—curved, 0.7 mm cup diameter, 14 cm.

#3 Scalpel blade holder.

#15 Scalpel blade.

#11 Scalpel blade.

Dumont 55 forceps.

Adson forceps.

Agricola retractor.

Vannas spring scissors—curved, sharp, 8.5 cm, 5 mm cutting edge (angled up).

Extra-fine Bonn scissors (straight).

Aspiration tip (blunted 22 G needle).

Aspiration bottle.

Fine cotton-tipped swabs.

Surgical gloves.

Gelfoam® sponge (Pfizer).

3-0 silk sutures.

Skin staples and stapler.

Nutri-Cal®.

Betadine® Microbicides.

70 % EtOH.

Optimmune animal eye ointment.

Animal hair clippers.

Syringes (3 ml).

Laboratory or surgical tape.

Gauze pads (2″ × 2″).

Heating pad or small-animal incubator.

2.6 Animals

10 Week-old (~150 g), F344 female rats (Harlan), maintained under standard animal care conditions (12:12-h light/dark cycle, lights coming on at 6 a.m., room temperature 21 °C), with food and water ad libitum. An approved IACUC protocol must be obtained prior to the procedures.

3 Methods

1. 1 h prior to surgery, place scaffold in saline to remove any preservative residues.
2. Anesthetize the rat by IM injection of the Ketamine/ Dexdomitor (67 mg/kg/0.5 mg/kg, respectively) cocktail (*see* **Note 1**).

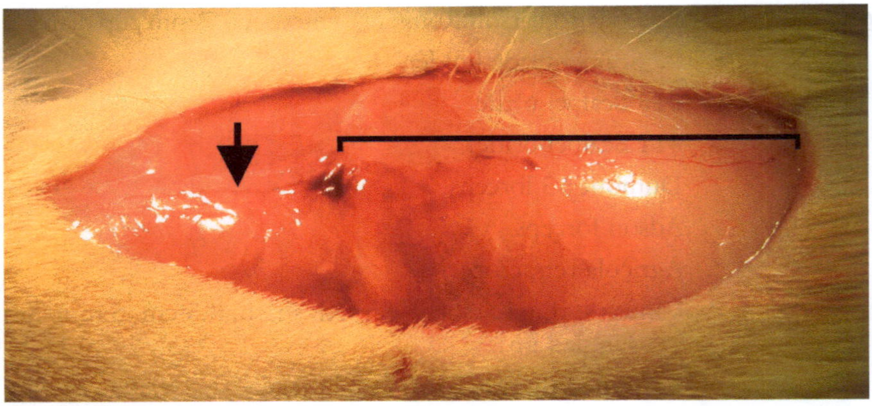

Fig. 1 Tissue revealed by the initial dorsal incision. The approximate midline is marked by the large blood vessel (*arrow*). Note the large fat pad (*bracket*). Rostral is to the *left* of the image

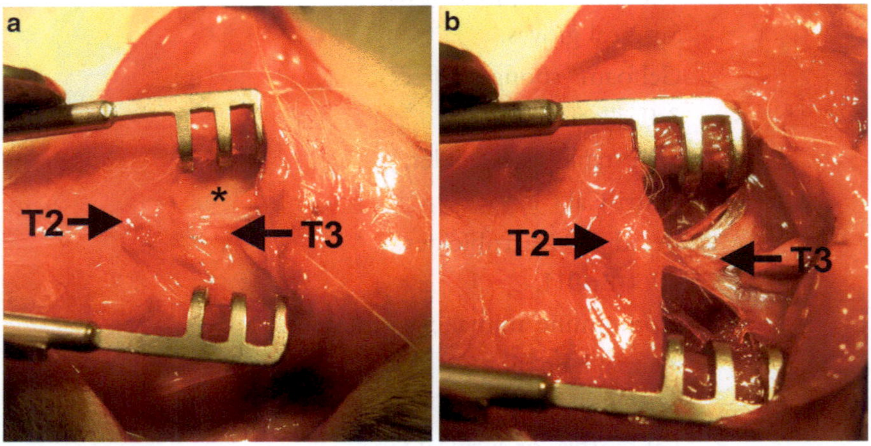

Fig. 2 (**a**) The view after separation of the superficial muscles. (**b**) The view after separation of the paraspinal muscles. Note the paraspinal muscles (*asterisk*) and T2 and T3 dorsal spinous processes (*arrows*)

3. Identify the T2 vertebra between the shoulders. The T2 dorsal spinous process is relatively large and protrudes from the spinal column, making it easy to identify by touch through the skin.
4. Shave the animal to expose the T1–T5 area.
5. Clean the surgery area using iodine and 70 % EtOH. Apply eye ointment.
6. Fix the animal to the surgical surface using tape. Raise the T3 area by inserting a few rolled gauze pads under the ventral side of the animal.
7. Open the skin from T1 to T5 using a #15 scalpel blade (Fig. 1).
8. Use a cotton-tipped swab to push the white fat pad (overlying T3–T5) caudally away from the surgery area (*see* **Note 2**).
9. Use a retractor and scissors to separate the superficial layer of muscles (Fig. 2).

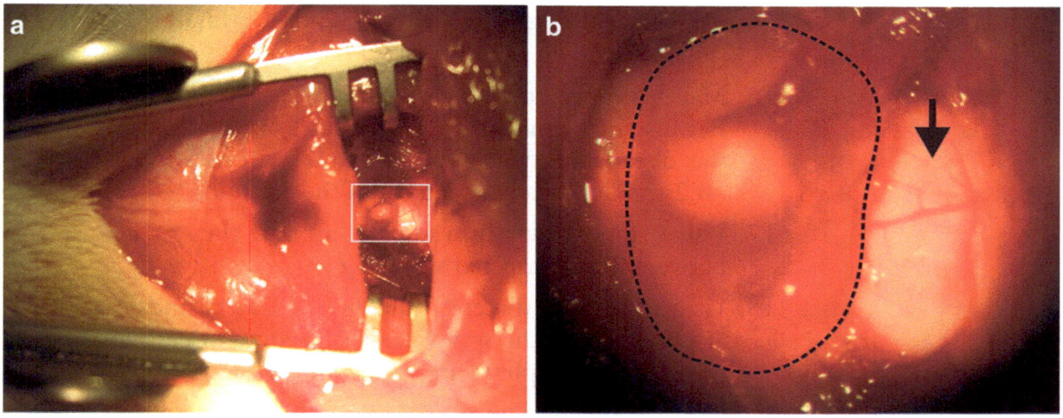

Fig. 3 The view after separation of the paraspinal muscles. Note T3 dorsal lamina (*box* in (**a**), *dashed line* in (**b**)) and the spinal cord (*arrow* in (**b**)) visible in the space between the T3 and T4 laminae

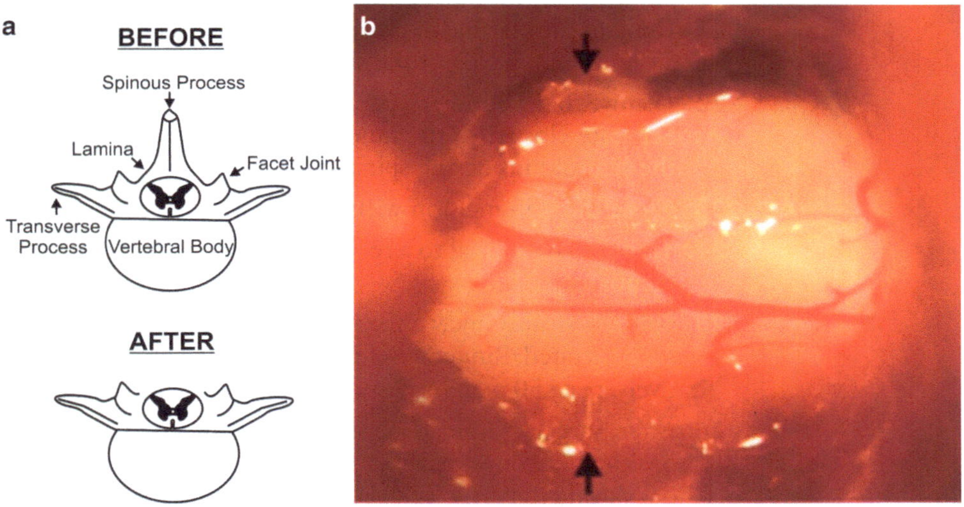

Fig. 4 (**a**) Schematic illustration of laminectomy. (**b**) T3 exposed spinal cord after laminectomy. *Arrows* indicate chipped bones at lateral edges of the laminectomy

10. Examine the deeper layer of muscles; identify the T3 lamina (Fig. 2a).
11. Use scissors to separate the deep paraspinal muscles from the T3 lamina. Lamina must be free of tissue (Figs. 2b and 3).
12. Reset the retractor to separate the deeper muscle layers.
13. Hold the T2 process with Adson forceps and perform a T3 laminectomy using rongeurs. Expose the spinal cord laterally up to just medial to the facet joints (Fig. 4, *see* **Note 3**).
14. Hold the dura with #55 Dumont forceps and make a longitudinal midline slit in the dura using a #11 scalpel blade.
15. Make a complete transection using the spring scissors. To preserve the dura, make two mediolateral hemisections of the spinal cord,

rather than a single cut across the entire width of the spinal cord (Fig. 5, *see* **Note 4**).

16. Make another complete transection as described in **step 15**. Because the spinal cord will retract from the transection sites, the distance between the two transections should be about ~0.2 mm less than the length of the scaffold.
17. Using standard vacuum suction, aspirate the spinal cord tissue between the transected sites (Fig. 6).
18. Using Dumont forceps, pull the dura away and examine the lateral edges of the spinal cord to verify transection completeness. Cut and aspirate any remaining spinal cord tissue.
19. Observe the bottom of the lesion site and make sure no tissue is left behind. Cut and aspirate any remaining tissue.
20. Wait for the bleeding to stop and gently remove blood clots using thin cotton-tipped swabs (*see* **Note 5**).
21. Implant the scaffold in the lesion site with the channels oriented along the long axis of the spinal cord (Fig. 7).
22. Using 3-0 silk suture, separately close the two layers of muscles. Separate closure of the inner and outer layer will help stabilize the scaffold inside the spinal cord.
23. Close the skin incision with staples.
24. Transfer the animal to a recovery area on a heating pad or in an incubator.
25. Subcutaneously inject 1 mg/kg Banamine, 20 mg/kg Ampicillin, and Lactated Ringer's Solution (0.3, 0.5, and 2.2 ml, respectively, per 150 g animal) and also inject 0.3 mg/kg Antisedan.
26. After surgery, animals should be examined daily for general health and wound healing. Bladders should be emptied twice a day during the first week [10]. This can subsequently be reduced to once a day as the animals recover. During the first 3 days of recovery, animals should receive a daily injection of the drugs as noted in the preceding step. Some animals may present blood in the urine in the following days; in this case additional daily ampicillin injections should be administered until recovery.
27. During the first week, both standard food and Nutri-Cal should be placed inside the cage, rather than in the wire rack of the cage lid, as animals may not be able to reach the wire rack.
28. Cages should be located on heating pads for the duration of the experiment. If this necessitates disconnection from an automatic water supply, provide water in bottles with long angled spouts, as animals will not be able to reach standard short spouts.
29. Tracing of specific axon groups penetrating the scaffold can be done later in the experiment with injections of BDA or CTB to the appropriate sites.

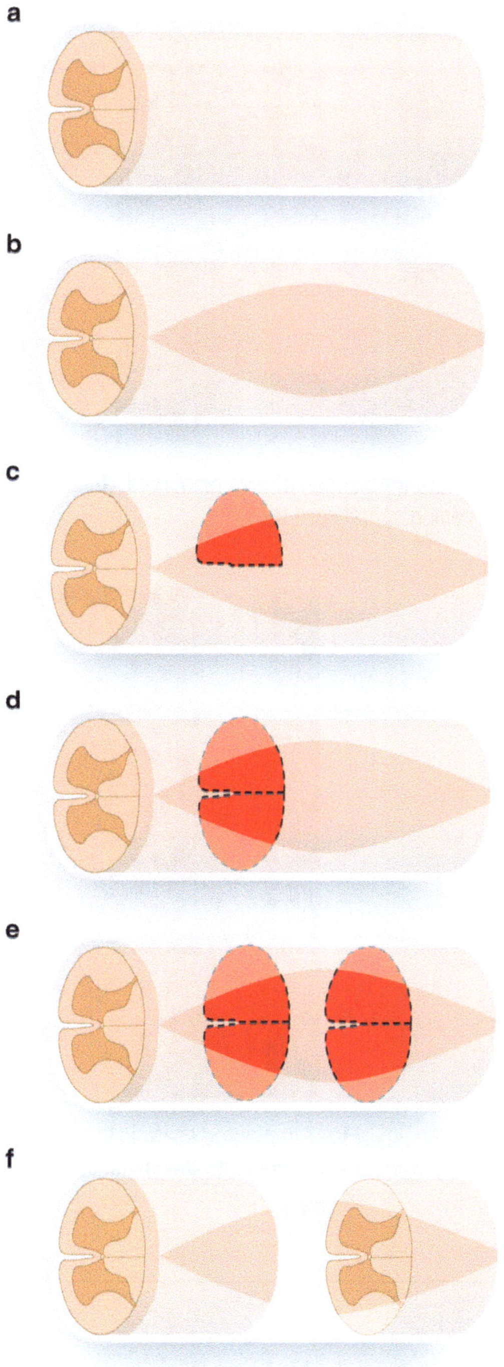

Fig. 5 Schematic illustration of complete transection. (**a**) Intact cord (*beige*) and dura (*white*). (**b**) Cut a longitudinal opening in the dura to allow access to the underlying spinal cord. (**c**) Make the first lateral hemisection (*red*) by passing scissors under preserved dura. (**d**) Repeat on the opposite side of the spinal cord to create a full transection. (**e**) Repeat steps (**c**) and (**d**) at a more caudal location (~0.2 mm less than the scaffold length to account for retraction). (**f**) Aspirate the tissue between the two cuts, leaving a gap the size of the scaffold

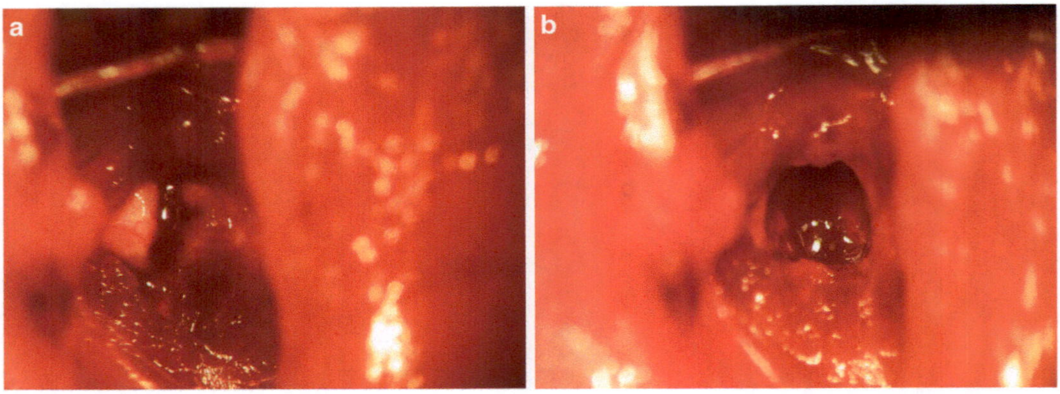

Fig. 6 (**a**) First complete transection of the spinal cord. (**b**) Lesion cavity after second complete transection and aspiration of intervening spinal cord tissue

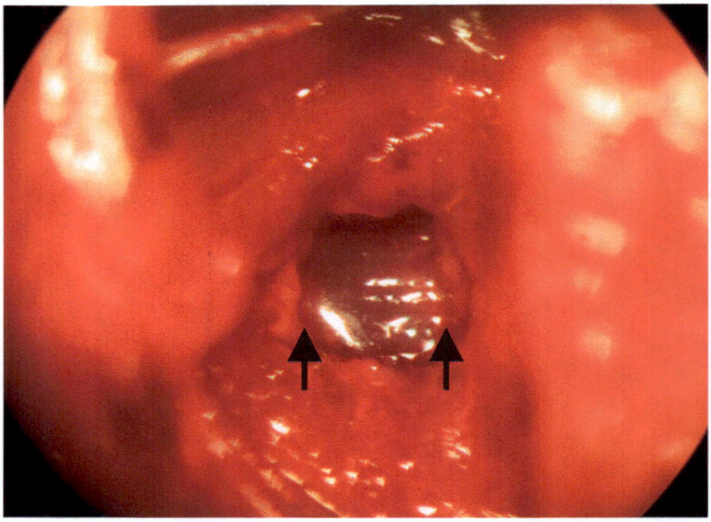

Fig. 7 Implanted scaffold in the lesion site. *Arrows* indicate the interface between the scaffold and the spinal cord stumps. Channels in the scaffold run horizontally in this image

4 Notes

1. Make sure animal is completely anesthetized by pinching one of its paws. Make sure no reaction is observed.
2. *It is important to be gentle with the fat pad* to minimize bleeding as it is highly vascularized. Do not attempt to remove it with a blade.
3. Do not remove the facet joints, as they are highly vascularized and are also critical for spinal column stability.

4. If there is bleeding, wait for it to stop (about 2 min), and then clean any blood clots. If you disturb the coagulation process too soon, bleeding will restart.
5. The duration and severity of bleeding varies from animal to animal, but it usually lasts approximately 5 min. If bleeding continues much longer, apply Gelfoam to the cavity. It will become red with blood within a minute. Wait another 2 min and carefully remove the Gelfoam. Pulling too hard on the Gelfoam will dislodge attached blood clots and may restart the bleeding.

Acknowledgements

This work is supported by NIH RO1-EBO14986, NIH R01-NS042291, NIH-NS067092, VA B7332R, Roman Reed Spinal Cord Injury Research Program of the State of California 10-274, the Dr. Miriam & Sheldon G. Adelson Medical Research Foundation, and the Bernard and Anne Spitzer Charitable Trust.

References

1. Alto LT, Havton LA, Conner JM, Hollis ER II, Blesch A, Tuszynski MH (2009) Chemotropic guidance facilitates axonal regeneration and synapse formation after spinal cord injury. Nat Neurosci 12(9):1106–1113
2. Gros T, Sakamoto JS, Blesch A, Havton LA, Tuszynski MH (2010) Regeneration of long-tract axons through sites of spinal cord injury using templated agarose scaffolds. Biomaterials 31(26):6719–6729
3. Jin Y, Tessler A, Fischer I, Houle JD (2000) Fibroblasts genetically modified to produce BDNF support regrowth of chronically injured serotonergic axons. Neurorehabil Neural Repair 14(4):311–317
4. Kwon BK, Liu J, Messerer C, Kobayashi NR, McGraw J, Oschipok L, Tetzlaff W (2002) Survival and regeneration of rubrospinal neurons 1 year after spinal cord injury. Proc Natl Acad Sci U S A 99(5):3246–3251
5. Lu P, Yang H, Jones LL, Filbin MT, Tuszynski MH (2004) Combinatorial therapy with neurotrophins and cAMP promotes axonal regeneration beyond sites of spinal cord injury. J Neurosci 24(28):6402–6409
6. Taylor L, Jones L, Tuszynski MH, Blesch A (2006) Neurotrophin-3 gradients established by lentiviral gene delivery promote short-distance axonal bridging beyond cellular grafts in the injured spinal cord. J Neurosci 26(38): 9713–9721
7. Stokols S, Tuszynski MH (2004) The fabrication and characterization of linearly oriented nerve guidance scaffolds for spinal cord injury. Biomaterials 25(27):5839–5846
8. Stokols S, Sakamoto J, Breckon C, Holt T, Weiss J, Tuszynski MH (2006) Templated agarose scaffolds support linear axonal regeneration. Tissue Eng 12(10):2777–2787
9. Basso DM, Beattie MS, Bresnahan JC (1995) A sensitive and reliable locomotor rating scale for open field testing in rats. J Neurotrauma 12(1):1–21
10. Gao M, Lu P, Bednark B, Lynam D, Conner JM, Sakamoto J, Tuszynski MH (2013) Templated agarose scaffolds for the support of motor axon regeneration into sites of complete spinal cord transection. Biomaterials 34(5): 1529–1536

Chapter 14

In Vivo Electroporation of Adult Mouse Sensory Neurons for Studying Peripheral Axon Regeneration

Saijilafu, Bo-Yin Zhang, and Feng-Quan Zhou

Abstract

Electroporation has been a widely used tool to introduce DNA plasmids or RNA oligos into cultured cells and recently in vivo into chick or mouse embryos. Here we report a rapid and efficient approach to transfect adult mouse dorsal root ganglion neurons in vivo with precise spatiotemporal control via electroporation. This approach will allow both gain- and loss-of-function experiments in vivo to study the function of adult sensory neurons, such as sensory axon regeneration.

Key words In vivo electroporation, Peripheral nervous system, Axon regeneration, Dorsal root ganglion, Adult mouse sensory neurons

1 Introduction

Genetic manipulation of adult neurons is particularly useful for studying genes that also play important roles during development because traditional knockout approach usually leads to early embryonic lethality or compensatory effects. The inducible transgenic approach, in which the gene of interest is acutely deleted in the adult stage, can significantly reduce the likelihood of compensatory gene expression changes or developmental defects. Another approach to specifically manipulate gene expression in adult neurons is viral vector-mediated gene targeting. However, both approaches involve time-consuming and expensive procedures, such as many rounds of animal crossing and tedious virus purification. Electroporation has become a routine and powerful technique to acutely manipulate gene expression in neurons both in vitro and in vivo. Gene delivery into embryonic neurons in vivo using electroporation [1] has greatly benefitted the study of gene functions during neural development. Similar approaches to deliver genes into certain postnatal or adult neurons have also been developed [2–4]. In axon regeneration studies, mouse dorsal root ganglion (DRG) neurons are becoming a valuable model [5].

Andrew J. Murray (ed.), *Axon Growth and Regeneration: Methods and Protocols*, Methods in Molecular Biology, vol. 1162, DOI 10.1007/978-1-4939-0777-9_14,

Each DRG neuron extends a single unipolar axon that splits into two branches: one branch extends to innervate the periphery targets (peripheral branch) and the other one goes up to enter the spinal cord (central branch). As a result, DRG neurons can be used for studying axon regeneration after either peripheral nerve or spinal cord injuries. To date, however, we know very little about the signaling pathways by which axon regeneration of DRG neurons is regulated in vivo because it requires genetic modulation of multiple genes simultaneously in adult DRG neurons. We recently have developed a novel in vivo electroporation technique that allows specific manipulation of multiple gene expression in adult DRG neurons simultaneously [6], thus providing a potential useful tool for in vivo dissection of pathways regulating mammalian axon regeneration. Using such approach, we have elucidated the important roles of several genes, including cytoskeletal proteins [7] and epigenetic modulators [8], in regulation of mouse sensory axon regeneration. Here, we provide a detailed description of such in vivo electroporation approach.

2 Materials

2.1 Animals and Surgery Materials

1. *Animals*: Adult female CF-1 mice (8–10 weeks old, weighing 30–35 g) from Charles River Breeding Laboratories are used for the experiments. The animals are housed in small cages with sawdust flooring and maintained in a controlled 12-h light–dark cycle. The animals have access to food and water ad libitum. All animal experiments should be performed in accordance with the animal protocol approved by the Institutional Animal Care and Use Committee.
2. *Stereotaxic apparatus*: Stereotaxic frame for rats (Model 900, David Kopf Instruments, UK) equipped with a mouse adaptor (Kopf Mouse Adaptor from 2Biol, Varese, Italy). We recommend placing the stereotaxic frame on a stable table to increase precision during surgery.
3. *Surgery materials*: 70 % alcohol, sterilized cotton swab sticks, sterilized gauze, surgery forceps, surgery scissors, bone rongeurs, small surgical retractors, small scalpel and scalpel blades, 30-G needles, 1.0-mL disposable syringes for intraperitoneal (i.p.) injection, fur remove clipper (miniARCO, Kent Scientific Corporation), and a surgery microscope (or a stereo dissection microscope).

2.2 Anesthetic Solution

Mix 1 mL ketamine (conc. 100 mg/mL) and 0.1 mL xylazine (conc. 100 mg/mL) with 8.9 mL normal saline (0.9 %) or phosphate-buffered saline (PBS). Store the mixed solution at 4 °C. Before each use, filter the solution with 0.22 μm syringe filters (Millex-HP, Millipore).

2.3 Plasmid Preparation

Plasmids are prepared using the commercially available Maxiprep Kit (e.g., Invitrogen Purelink) according to the manufacturer's protocol. Plasmid DNAs (e.g., pCMV-EGFP-N1) are diluted in the sterile, distilled, and deionized water with a final concentration of 2.0–4.0 μg/μL. To help visualize the DNA/RNA solution, add 0.01–0.02% fast green dye (from 1 % stock).

2.4 Microinjection and Electroporation Equipment

The glass micropipette used for the microinjection is made by pulling the capillary glass (10 cm, standard wall, Sutter Instrument) with a micropipette puller (e.g., Sutter Instrument). A microelectrode holder (with port for air) is used to fix the micropipette. The Picospritzer II (Parker Inc.) is used for the injection. The BTX ECM830 Square Wave Electroporation System is used for the electroporation. The electrode is a tweezer-like electrode (Ø1.2 mm) custom made by Harvard Apparatus (*see* ref. [6]).

2.5 Preparation of Microinjection Glass Pipette

1. Pull the capillary glass tube using the micropipette puller.
2. Pinch off the tip of the pulled capillary glass pipette with jeweler's #5 forceps to generate an opening of about 80 μm in external diameter (*see* **Note 1**).
3. Sterilize the glass pipette under UV light on a clean bench for about 20–30 min.

2.6 Perfusion Solution

Add 40 g of paraformaldehyde to 1000 mL of PBS and heat to 60 °C while stirring to dissolve the paraformaldehyde. Cool off the solution and store at 4 °C before use.

3 Method

3.1 Neurosurgery to Expose the Dorsal Root Ganglion

1. Anesthetize the mouse by intraperitoneal injection of the previously prepared anesthetic solution of ketamine (100 mg/kg) and xylazine (10 mg/kg).
2. Shave the animal's back fur using the fur clipper, and disinfect the back skin with 70 % alcohol (Fig. 1).
3. Put the mouse on heated blanket (35.0 °C), and monitor the rectal temperature with a temperature probe.
4. To test the depth of animal's consciousness, pinch the toes and tail and observe the animal for a response.
5. Fix the animal on the stereotaxic frame and in the mouse adapter.
6. Make a 3-cm-long skin incision in the midline of the back using a sharp scalpel blade, following the rostrocaudal direction.
7. Dissect free the paraspinous muscles, such as the musculus multifidus and the musculus longissimus lumborum, from the L4–L5 spinous processes, and expose the transverse processes by scraping off attached connective tissue (Fig. 2) (*see* **Note 2**).

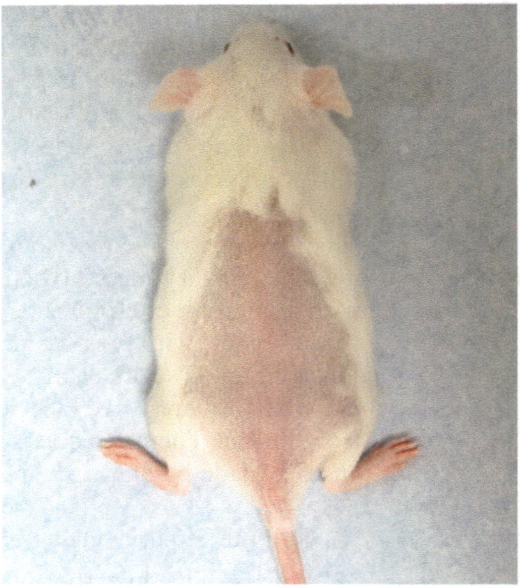

Fig. 1 Preparation of the mouse for surgery. After anesthetization, the back fur of the mouse is shaved and the exposed skin is disinfected

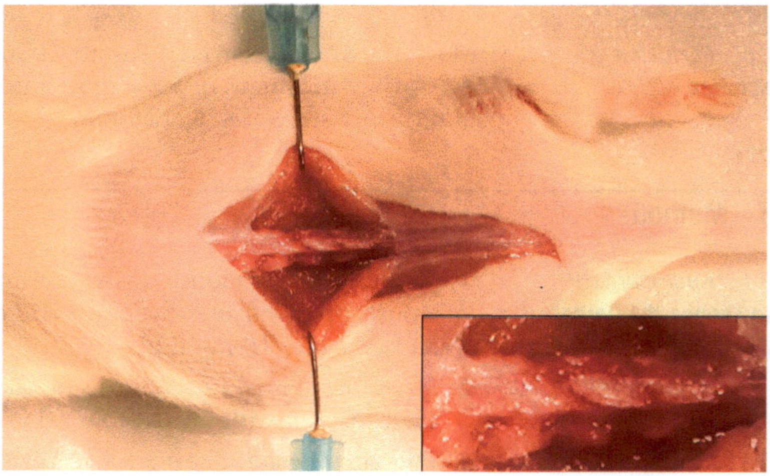

Fig. 2 Surgical exposure of the vertebrate. The back muscle is dissected to expose the L4–L5 vertebrate

8. Remove the processus accessorius and parts of the processus transversus under a surgery microscope with micro-rongeurs to visualize the dorsal side of the left L4 and L5 DRGs (*see* **Note 3**).
9. Stop any bleeding with sterilized cotton swab sticks and gauze (*see* **Note 4**).

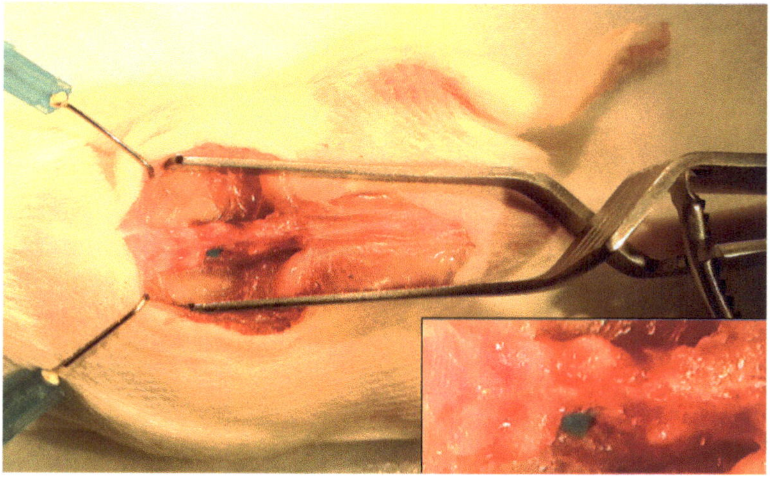

Fig. 3 Microinjection into the dorsal root ganglion. 1.0 μL solution of DNA plasmids or RNA oligos is gradually injected into the DRG using the air pressure generated by the Picospritzer. Note the round DRG filled with the fast green dye

3.2 Microinjection

1. Load the DNA plasmid or RNA oligo solution (2–4 μL) into the pulled capillary micropipette with a micropipette loading tip.
2. Fix the micropipette onto a microelectrode holder (with port for air), and connect it to the Picospritzer.
3. Under the surgery microscope, carefully insert the tip of the micropipette into the DRG with a 45° angle (*see* **Note 5**).
4. Gradually inject 1.0 μL solution of DNA plasmids or RNA oligos into the DRG using the Picospritzer (pressure: 30 psi, duration: 8 ms) (Fig. 3) (*see* **Notes 6** and 7).

3.3 In Vivo Electroporation

1. Add a few drops of PBS solution on the surface of the electrode.
2. Clean up any bleeding with sterilized cotton swab sticks, and fill the surgery site with sterile saline to prevent desiccation.
3. Gently pinch the target DRG with the electrode, and apply five square electric pulses with the BTX ECM830 system (15-ms pulses at 35 V with 950-ms interval) (Fig. 4) (*see* **Note 8**).
4. Carefully close the muscle and skin separately with 5-0 nylon sutures. Place the mouse on a heated blanket (35.0 °C) until the animal completely recovers from anesthesia.

3.4 Sciatic Nerve Crush

1. A few days (depending on the experimental design) after the electroporation, anesthetize the mouse again with intraperitoneal injection of the anesthetic solution of ketamine (100 mg/kg) and xylazine (10 mg/kg).
2. Expose the sciatic nerve at the sciatic notch by a small incision.

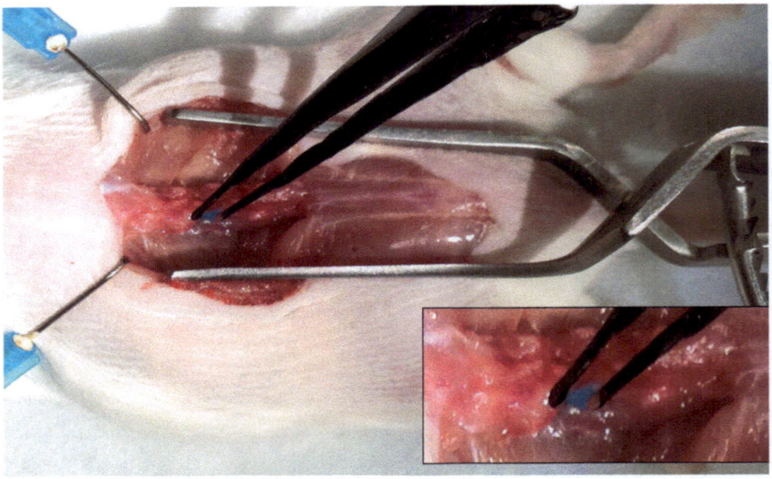

Fig. 4 In vivo electroporation. The DRG is placed between the two tips of the tweezers like electrodes, and the electric pulses are then applied

3. Crush the nerve with jeweler's #5 forceps for 15 s twice, and mark the crush site with a 10-0 nylon epineural suture (*see* **Note 9**).
4. Close the skin with 5-0 nylon sutures.
5. Place the mouse on a heated blanket (35.0 °C) until the animal completely recovers from anesthesia.

3.5 DRG and Sciatic Nerve Harvest

1. A few days after the sciatic nerve crush (depending on the experimental design) terminally anesthetize the mouse with anesthetic solution of ketamine (100 mg/kg) and xylazine (10 mg/kg).
2. Transcardially perfuse the mouse with PBS (pH 7.4) followed by ice-cold 4 % paraformaldehyde (PFA) in PBS.
3. Under the dissecting microscope, carefully dissect out the DRG together with the whole sciatic nerve and post-fix the tissues overnight in 4 % PFA at 4 °C.

3.6 Imaging and Measuring the Fluorescence-Labeled Sensory Axons

1. Place the DRG or the sciatic nerve between two glass slides, and flatten the whole mounted tissue with pressure.
2. Place the flattened tissues under an automated epifluorescent microscope equipped with the mosaic acquisition and processing capacity (*see* **Note 10**). An alternative approach is to take multiple overlapping images manually and stitch the images together using the Mosaic plug-in of the ImageJ program.

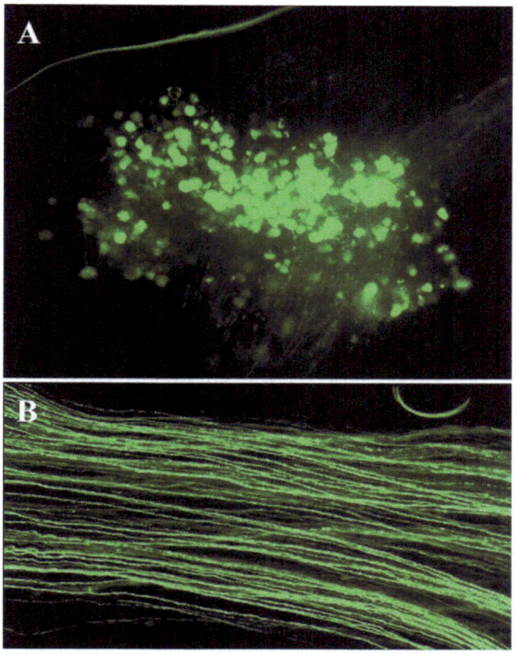

Fig. 5 Representative result of in vivo electroporation. Efficient transfection and expression of EGFP can be achieved in adult DRG neurons (**a**) and sciatic nerve axons (**b**)

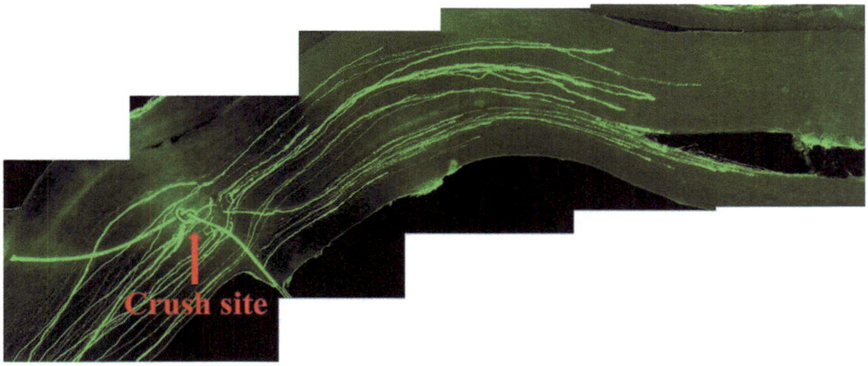

Fig. 6 Study peripheral axon regeneration using in vivo electroporation. Successful axon regeneration can be directly observed by labeling regenerating axons with EGFP. An epineural suture is used to label the nerve crush site

3. Take the images of the DRG (Fig. 5a) or the sciatic nerve (Fig. 5b) with a CCD camera, and assemble the images.
4. Manually trace all identifiable fluorescent-labeled axons in the sciatic nerve from the crush site (marked with the 10-0 epineural suture) to the distal growth cone to measure the length of regenerating axons (Fig. 6).

4 Notes

1. When making the micropipette for injection, pinch the tip of pulled capillary glass pipette with a 30–45° angle to get a sharp end, which makes it easier for the micropipette to penetrate into the DRG.
2. For accurate identification of the L4/L5 DRG, the superior edge of the iliac crest is a very useful anatomic landmark. Usually, the transverse process of L5 vertebrate is located around the superior edge of the iliac crest.
3. When surgically exposing the DRG, the processus accessorius and parts of the processus transversus need to be carefully removed with the micro-rongeurs to avoid any damage to the spinal cord, DRG, and nerve tissue.
4. Before microinjection, it is very important to stop any bleeding with cotton swab sticks, as the bleeding can affect precision of the microinjection.
5. During microinjection, insert the tip of the glass micropipette into the DRG with a 45° angle, which will help the diffusion of the injected solution inside the DRG.
6. The successful microinjection is the key point of the entire procedure. To increase the precision of the microinjection, add fast green dye into the DNA/RNA solution to visualize the injection result. A successful injection often shows a clearly defined round-shaped DRG.
7. During microinjection, if the injected solution (colored with fast green) is seen in other regions, adjust the position of the pipette tip by pulling back or penetrating a little further until the solution can be injected into the DRG. Usually it takes multiple injections at different sites to fill the DRG.
8. Make sure that the contact surfaces of the electrode and DRG are filled with PBS, which will increase transduction of the electric current. A successful electroporation usually results in generation of gas bubbles around the DRG tissue.
9. For analyzing sciatic nerve regeneration, accurate identification of the crush site is crucial. Sometimes the epineural suture marking the crush site may fall off during the dissection process to remove the sciatic nerve. It is important to remark the crush site with suture as soon as possible while the crush site can still be identified by other visible markers.
10. Although the flattened nerve can be imaged by the epifluorescent microscope, some axons deep in the tissue may not be identified. Flipping the slide to image both sides of the nerve may help identify more labeled axons. Alternatively, confocal microscopy may be used to generate a 3D reconstruction of the nerve.

Acknowledgments

This work was supported by grants to F.Q.Z. from NIH (R01NS064288) and the Craig H. Neilsen Foundation.

References

1. Saito T, Nakatsuji N (2001) Efficient gene transfer into the embryonic mouse brain using in vivo electroporation. Dev Biol 240(1):237–246
2. Matsuda T, Cepko CL (2007) Controlled expression of transgenes introduced by in vivo electroporation. Proc Natl Acad Sci U S A 104(3): 1027–1032
3. Barnabe-Heider F et al (2008) Genetic manipulation of adult mouse neurogenic niches by in vivo electroporation. Nat Methods 5(2): 189–196
4. Konishi Y, Stegmuller J, Matsuda T, Bonni S, Bonni A (2004) Cdh1-APC controls axonal growth and patterning in the mammalian brain. Science 303(5660):1026–1030
5. Saijilafu, Zhang BY, Zhou FQ (2013) Signaling pathways that regulate axon regeneration. Neurosci Bull 29(4):411–420
6. Saijilafu, Hur EM, Zhou FQ (2011) Genetic dissection of axon regeneration via in vivo electroporation of adult mouse sensory neurons. Nat Commun 2:543
7. Hur EM et al (2011) GSK3 controls axon growth via CLASP-mediated regulation of growth cone microtubules. Genes Dev 25(18): 1968–1981
8. Liu CM et al (2013) MicroRNA-138 and SIRT1 form a mutual negative feedback loop to regulate mammalian axon regeneration. Genes Dev 27(13):1473–1483

Part V

Assessing the Extent of Axon Regeneration

Chapter 15

Assessing Motor Outcome and Functional Recovery Following Nerve Injury

Ronald Deumens, Claudia Marinangeli, Ahmet Bozkurt, and Gary Anthony Brook

Abstract

Behavioral tests significantly contribute to our understanding of nerve function after experimental lesions and/or therapeutic intervention. In particular, the rat sciatic nerve has proven to be a valuable animal model to study nerve injury and repair. Here, we describe how to optimally use the commercially available CatWalk system to obtain a detailed and objective analysis of dynamic and static gait parameters.

Key words Gait analysis, Peripheral nerve injury, Repair, Nerve lesions, Locomotion, Behavioral tests, Functional recovery, Rodents

1 Introduction

Behavioral tests are a major asset in the field of experimental nerve injury and repair, giving meaningful information about the functionality of a peripheral nerve. Within this scientific domain, the sciatic nerve is one of the most frequently used nerves in rodent models. Sciatic nerve injury can substantially affect behavioral outcome. Firstly, somatosensory and motor functions can be negatively influenced ("loss of function") in the affected hind limb due to an impaired signal conduction along the sciatic nerve. Secondly, nerve injury is a frequent cause of pain with a major role for injury-evoked spontaneous electrical discharges generated in ectopic locations (such as at the site of nerve injury or at the dorsal root ganglion) [1, 2] and that drive processes of sensitization ("gain of function"). Thirdly, injury to a peripheral nerve can also lead to perturbations of other higher brain functions such as cognition and mood. All these nerve injury-evoked processes can potentially affect the behavior of the animal. Therefore, it is of pivotal importance that behavioral tests for the assessment of the functionality

Andrew J. Murray (ed.), *Axon Growth and Regeneration: Methods and Protocols*, Methods in Molecular Biology, vol. 1162, DOI 10.1007/978-1-4939-0777-9_15, © Springer Science+Business Media New York 2014

of a particular nerve are validated. This means that a behavioral parameter shows a clear deficit upon nerve injury and a recovery upon nerve regeneration. The classical method to assess sciatic nerve functionality is a form of footprint analysis, termed the sciatic function index (SFI), and was developed in the early 1980s [3]. In naive, uninjured conditions, the SFI has a value around 0, but this value drops to around −100 when the sciatic nerve is completely injured (transection or neurotmesis). After sciatic nerve crush (or axonotmesis), an injury model which is characterized by almost complete axon regeneration [4], a gradual return towards preoperative values is observed. In our studies, we have validated the use of a novel behavioral test for sciatic nerve injury in the rat. This test uses an apparatus, the CatWalk, which is commercially available and has been developed for assessment of gait in rodent models of spinal cord injury [5]. The CatWalk uses light captured within a glass runway. When an object (e.g., a rodent paw) comes into contact with the glass surface the light is deflected downwards and detected by a video camera positioned underneath the glass runway. During the animal's crossing of the runway, highlighted paw prints are detected and the information about the paw prints is then sent to a computer equipped with CatWalk software to analyze a wide range of parameters. These parameters include individual paw parameters such as paw print size, paw print intensity, stance phase (i.e., average duration that paw is in contact with surface), swing phase (i.e., average duration that the paw is not in contact with surface), and stride length (i.e., the distance between two subsequent steps made by the same paw). Furthermore, parameters which require information about more than one paw, for example regularity of paw placing (the sequence at which the four paws are placed during gait), can also be determined. We observed that complete sciatic nerve injury (transection) caused rapid and persistent deficits in a range of parameters including hind paw base of support (i.e., distance between hind paws), paw print width, length, area and intensity, as well as stance and swing phase [6]. Although no difference was found for the regularity of paw positioning, the specific sequence of paw placements changed after sciatic nerve injury and also the timing of the affected hind paw was delayed [6]. When crushing the rat sciatic nerve instead of transecting it, injury-induced deficits in gait parameters gradually recovered over the course of a number of weeks. Some parameters reached preoperative values (e.g., stance phase, swing phase), while others did not recover fully (e.g., paw print width, length, and area) [7]. To date, a number of studies have successfully used the CatWalk in the assessment of rodent sciatic nerve repair [8].

2 Materials

2.1 CatWalk Apparatus and Computer, Equipped with CatWalk Software

To be purchased from Noldus Information Technology, The Netherlands. The device consists of an elevated glass walkway of a length of approximately 1.5 m and a width that can be adapted (typical width around 10 cm). Moreover, it is equipped with a high-speed color camera (positioned underneath the walkway) for recording of paw prints with accurate spatial and temporal resolution.

2.2 Experimental Room

A darkened room is advisable, although the newer CatWalk systems are better adapted for use in rooms which are not completely dark.

2.3 Motivational Cues: Food Rewards

The ultimate level of performance of rats on the CatWalk will depend on their motivation to execute the test, making consecutive, uninterrupted runs. This motivation can be reinforced by using food rewards at the end of the runway. In our hands, 45 mg sugar pellets (Noyes Precision pellets PJPPP-0045; Sandown Chemical Ltd., Hampton, UK) have proved to be attractive motivational cues. The rats need to be habituated to the food rewards, and hereto we have regularly (i.e., daily) provided such rewards to the cages of the animals in limited amounts.

2.4 Animals

Selection of the animal strain can greatly influence study outcomes. We found that different rat strains (e.g., Lewis, Wistar, Sprague Dawley) showed substantial differences in gait in naive, uninjured conditions. In order to reliably detect nerve injury-evoked alterations in some gait parameters, thereby creating windows of opportunity for altered gait parameters, the choice of rat strain needs to be considered beforehand (for further explanation, *see* Subheading 3.4). Furthermore, the choice of rat strain and sex as well as lesion severity have been reported to have a substantial influence on automutilation [9–12]. The loss of any digits from the animals under investigation would have a detrimental effect on any footprint-based analysis. In this context, Lewis rats have proved to be the most reliable strain for such studies.

3 Methods

3.1 Training Period

1. To obtain a stable performance on the CatWalk, rats need to be trained, meaning that they need to reach a level in which the animals show a stable gait and no hesitation during several consecutive runs (*see* **Note 1**). In our hands, a combined protocol of food restriction (e.g., 10–15 g of food per rat per day) and regular access to food rewards (e.g., sugar pellets) has proved successful. These latter rewards are used at the farther

end of the CatWalk runway to motivate the animals. The training period typically lasts for 2 weeks, during which the rats are trained for 5 days per week (*see* **Note 2**).

2. Place the food rewards at one end of the glass runway.
3. Take the rats from their home cage (*see* **Note 3**).
4. Place the rat on the glass runway at the end opposite to that containing the food rewards (*see* **Note 4**).
5. Allow the rat to freely move to the other end of the glass runway before gently removing it from the CatWalk. If the rat attempts to turn around, actively prevent this by gently using your hands as barriers. The rat has to learn to traverse the runway without turning. For the first crossings, rats may have the tendency to run away from the experimenter, which can frequently lead to uninterrupted runs to the other side of the runway. This behavior does not mean that the rat has acquired a stable performance. Instead, for the following runs, when rats familiarize with the procedure, they will start demonstrating exploratory behavior within the runway, leading to (frequent) interruptions during the crossings. The training period is required until the rats reach a state in which they will make consecutive, uninterrupted crossings (*see* **Note 5**).
6. Use a minimum of three runway crossings per day for a period of 2 weeks. During the first days, rats typically do not eat the food rewards at the farther end of the runway, most likely because they have to "learn" to find the food pellets at this location in the "still relatively new" environment. After a couple of days (typically ~5), rats will have learned where to find their food rewards, leading to a substantial reduction in the number of interruptions during the crossings. Situations in which the food rewards are not found to be sufficiently interesting for the animal result in suboptimal training (*see* **Note 6**). Therefore, the motivational value of the food reward is an important aspect of CatWalk training. After a minimum of three CatWalk crossings, the animal can be returned to its home cage (*see* **Note 7**).
7. If a rat does not make uninterrupted runs close to the end of the second training week, training needs to be intensified. This means that a session of three CatWalk crossings will be repeated several times. In between such sessions it is important to respect a certain amount of "recovering time", meaning that the animal will be housed in its home cage for a duration of, e.g., 10 min and then trained again. The essential aspect is that the animal will not be distracted by environmental cues (such as the CatWalk side barriers) and consistently walk towards the food rewards at the end of the runway (*see* **Note 8**).

3.2 CatWalk Preparation for the Experiment

Before recording any crossings, it is important to ensure a proper calibration of the system. Also, the camera should be optimally focused on the runway, which can be easily verified, e.g., using a small paper with letters on the CatWalk runway and validating a clear visual of the letters. The configuration settings should be set to levels that do not lead to detection of background signals but give a sensitive detection of the paw prints during a CatWalk crossing. Identical settings have to be used throughout the experiment, and therefore the configuration settings should be saved in a configuration file before taking baseline recordings. This configuration file is then loaded for each recording day. It is furthermore recommended to clean the runway between one animal and the next.

3.3 Baseline Recording of Gait

1. Only use rats which have acquired sufficient training to make consecutive, uninterrupted crossings for baseline recordings.
2. Take baseline recordings according to the instructions of the supplier (Noldus Information Technology, The Netherlands).
3. Select runs with proper crossing durations. The velocity at which the animals traverse the runway is one of the most important factors to control because it affects many gait parameters irrespective of nerve lesions [13]. Obviously, the CatWalk apparatus does not make rats walk at a fixed speed, but the crossing duration is recorded and the investigator should select only those crossings for analysis that fall within an acceptable time frame. For a runway distance of about 1.5 m, a crossing time of between 1 and 2 s is recommended.
4. Select runs with at least three step cycles. A step cycle is a sequence of steps in which each paw is used/detected once. With three step cycles, parameters are calculated as averages of three values per paw. Typically, we have used a minimum of three runs per animal, but this is not a strict rule. The more runs being used for analysis, the more reliable the data will be.

3.4 Postsurgical Low-Dose Training and Recording of Gait

1. To obtain postsurgical recordings at the times relevant for your particular investigation, it is important to maintain a low dose of training during the postsurgical period. This means that the animal should be allowed to make at least five runway crossings for at least 2 days per week. If postsurgical recordings are done at this interval or more frequently, no additional training sessions should be needed. However, animal behavior never fits exact rules or protocols, which means that one should pay attention to the performance level of each individual animal. In the unfortunate case that an animal no longer makes consecutive uninterrupted crossings, a low dose of additional training should be considered. Patience is required, and animals should be exposed to repetitive training sessions at approximately 10-min intervals. In the postsurgical phase, the rat

normally does not need to be maintained on a restricted food protocol. Instead, the evening before the training or the testing day, the rat should be completely deprived of food and placed on ad libitum access to food immediately after recording the crossings. Typically, we have used a minimum of three runs per animal per postsurgical time point of investigation.

2. Similar to the criteria for baseline recordings, postsurgical recordings should only be included if the crossing time is between 1 and 2 s (for a runway distance of approximately 1.5 m). Use clean runways at the start of postoperative time point recording for each animal.

3.5 Analysis of Gait Parameters

1. Analysis of runs is performed using CatWalk software. Depending on the version used, paw identification is done manually or automatically (*see* **Note 9**).
2. In the first days after sciatic nerve lesion (transection or crush of the entire sciatic nerve trunk), the intensity of the paw prints is particularly low, rendering very feeble prints and perhaps below the detection threshold of the software. This implies that one can analyze some but not other parameters. For instance, parameters requiring more than one paw (e.g., regularity of paw placements) cannot be used as there is no detection of the affected hind paw. Use of individual paw parameters such as print width, length, and area or intensity can be set to 0, but this should be explicitly mentioned in the description of research methods.
3. Sciatic nerve lesions impair the rodent's use of the hind limb, leading to detection of predominantly the heel region of the paw on the glass runway. As such, parameters like paw print width, length, and area will be reduced (*see* **Note 10**).
4. The choice of rat strain can affect the outcome of gait parameters after nerve injury. Sciatic nerve injury affects coordination-related parameters. Although step sequence patterns do not become significantly irregular, the specific patterns used may change. Six normal step sequence patterns have been described in rats [14] (Fig. 1). We have previously found that naive male Wistar rats show a nearly equal use of the two cruciate (fore paw side A–fore paw side B–hind paw side A–hind paw side B) and two alternate patterns (fore paw side A–hind paw side A–fore paw side B–hind paw side B). After sciatic nerve transection, the rats nearly exclusively use one cruciate and one alternate pattern, most likely to reduce the burden on the affected hind paw [6]. These results are very different when using female Lewis rats, which already show a major preference to one alternate pattern at baseline [7]. Sciatic nerve injury in these animals will therefore have much more subtle effects on such step sequence patterns.

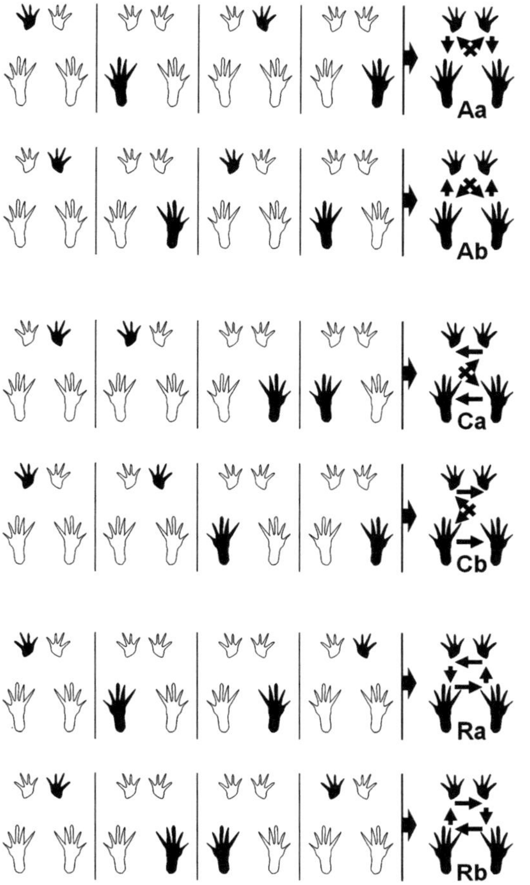

Fig. 1 Schematic representation of normal step sequence patterns as described in ref. [14]. These patterns can be alternating between the right and left sides of the body (Aa and Ab); they can cruciate each time a paw is placed (Ca and Cb) or rotate (Ra and Rb). There may be a preference for some of these patterns under naive, uninjured conditions, depending on the animal strain. Since a nerve injury will favor the use of particular patterns, selecting the proper animal strain is important to ensure an injury-induced effect on normal step sequence patterns; *see* ref. [7]

4 Notes

1. In contrast to rats, mice are not trained as these animals do not acquire the skill to make uninterrupted runs. Instead, mice are simply allowed to move freely on the CatWalk runway and runs are recorded, based on the anticipation of the mice producing uninterrupted runs during the test. This is obviously a more time-consuming process in comparison to recording runs from trained rats.
2. Animal handling: Every training day, the rat ought to be handled gently before its training session. This implies holding,

caressing, etc. of the animal by the experimenter for about 5 min, which helps to get the animal and experimenter "used to one another." At any time, the animal is held in a gentle way and never at the tail.

3. Avoid moist paws as this can substantially affect many CatWalk parameters, which rely on light reflections in the CatWalk system. Ensuring regular cleaning of the CatWalk and the animals' cages will avoid moist paws.
4. The animal is placed for a few seconds in an empty, clean cage without lid before being transferred to the CatWalk runway. The purpose of this action is to ensure that any dirt (e.g., bedding material sticking to the animal) will not be transferred onto the CatWalk runway when the animal is placed on it. Although this process takes a little more time, it saves time in the long run by reducing the "background" signals caused by a dirty runway.
5. The newer CatWalk systems are equipped with covering boxes containing a red-light source. During the early stages of the training in which the animals still show exploratory behavior with regular turning behavior, the covering boxes should not be placed to close the top of the CatWalk system. As such, the experimenter will have continuous access to the animal so as to actively prevent turning behavior. Once the rat has acquired a good level of performance, the cover box can be lowered onto the CatWalk system, thereby closing the system at the top, and be used as such during the last training day(s) and during crossing that will be recorded (e.g., baseline or postoperative time points).
6. During the first days of training, before taking the animal off the runway it is essential to allow some time for the rat to explore the farther end of the CatWalk, where the food rewards are located. By this approach, the animal is allowed to find out where the food rewards are located. Then, the rat can be placed back either in the empty, clean cage if more training runs will follow or back in its home cage if the training session is over. As mentioned above, the training effect largely relies on the motivation of the animal for obtaining food rewards. For this reason, it is important to ensure the attractive nature of the food rewards before starting CatWalk training.
7. Keep the animals at an identical light–dark cycle and use temperature-controlled conditions throughout the experiment. Any change in the biorhythm of the animal can negatively influence its behavior, including performance on the CatWalk.
8. Distractions negatively influence the training effect. For this reason, it may help to clean the runway (e.g., using alcohol and water) between training sessions of different animals. As such odors derived from animals that were trained previously are reduced and form less of an interference cue.

9. Analysis of CatWalk parameters depends on the software version. In older versions, paw prints need to be manually identified (fore paw right and left, hind paw right and left); in newer versions the CatWalk software has automatic paw identification, but this identification should be verified by the investigator as the system is not error proof. In fact, when animals have the propensity to place their hind paws practically at the same location as the fore paws and when the time between fore paw lift off and hind paw placement is very small, the software tends to erroneously maintain the same paw identity. Each time this happens, the identification process needs to be manually corrected.
10. All measured CatWalk data may not be solely dedicated to sciatic nerve functionality. For instance, pain hypersensitivity after nerve injury might also have an influence on gait parameters to a certain extent, as suggested by some authors [15]. Obviously, such effects cannot be prevented and are integral parts of the investigation. However, authors should be aware of this possible mechanism when describing their findings in manuscripts.

Acknowledgements

This work was supported by an FSR Incoming Postdoctoral Fellowship of the Académie Universitaire "Louvain", co-funded by the Marie Curie Actions of the European Commision (to R.D.).

References

1. Devor M (2009) Ectopic discharge in Abeta afferents as a source of neuropathic pain. Exp Brain Res 196:115–128
2. Berger JV, Knaepen L, Janssen SP et al (2011) Cellular and molecular insights into neuropathy-induced pain hypersensitivity for mechanism-based treatment approaches. Brain Res Rev 67:282–310
3. de Medinaceli L, Freed WJ, Wyatt RJ (1982) An index of the functional condition of rat sciatic nerve based on measurements made from walking tracks. Exp Neurol 77:634–643
4. Luis AL, Amado S, Geuna S et al (2007) Long-term functional and morphological assessment of a standardized rat sciatic nerve crush injury with a non-serrated clamp. J Neurosci Methods 163:92–104
5. Hamers FP, Lankhorst AJ, van Laar TJ et al (2001) Automated quantitative gait analysis during overground locomotion in the rat: its application to spinal cord contusion and transection injuries. J Neurotrauma 18:187–201
6. Deumens R, Jaken RJ, Marcus MA et al (2007) The CatWalk gait analysis in assessment of both dynamic and static gait changes after adult rat sciatic nerve resection. J Neurosci Methods 164:120–130
7. Bozkurt A, Deumens R, Scheffel J et al (2008) CatWalk gait analysis in assessment of functional recovery after sciatic nerve injury. J Neurosci Methods 173:91–98
8. Bozkurt A, Scheffel J, Brook GA et al (2011) Aspects of static and dynamic motor function in peripheral nerve regeneration: SSI and CatWalk gait analysis. Behav Brain Res 219: 55–62
9. Wall PD, Devor M, Inbal R et al (1979) Autotomy following peripheral nerve lesions: experimental anaesthesia dolorosa. Pain 7: 103–111

10. Carr MM, Best TJ, Mackinnon SE et al (1992) Strain differences in autotomy in rats undergoing sciatic nerve transection or repair. Ann Plast Surg 28:538–544
11. Weber RA, Proctor WH, Warner MR et al (1993) Autotomy and the sciatic functional index. Microsurgery 14:323–327
12. Wang H, Sorenson EJ, Spinner RJ et al (2008) Electrophysiologic findings and grip strength after nerve injuries in the rat forelimb. Muscle Nerve 38:1254–1265
13. Koopmans GC, Deumens R, Brook G et al (2007) Strain and locomotor speed affect overground locomotion in intact rats. Physiol Behav 92:993–1001
14. Cheng H, Almstrom S, Gimenez-Llort L et al (1997) Gait analysis of adult paraplegic rats after spinal cord repair. Exp Neurol 148:544–557
15. Vrinten DH, Hamers FF (2003) 'CatWalk' automated quantitative gait analysis as a novel method to assess mechanical allodynia in the rat; a comparison with von Frey testing. Pain 102:203–209

Chapter 16

The Use of an Adeno-Associated Viral Vector for Efficient Bicistronic Expression of Two Genes in the Central Nervous System

Thomas Haynes Hutson, Claudia Kathe, Sean Christopher Menezes, Marie-Claire Rooney, Hansruedi Bueler, and Lawrence David Falcon Moon

Abstract

Recombinant adeno-associated viral (AAV) vectors are one of the most promising therapeutic delivery systems for gene therapy to the central nervous system (CNS). Preclinical testing of novel gene therapies requires the careful design and production of AAV vectors and their successful application in a model of CNS injury. One major limitation of AAV vectors is their limited packaging capacity (<5 kb) making the co-expression of two genes (e.g., from two promoters) difficult. An internal ribosomal entry site has been used to express two genes: However, the second transgene is often expressed at lower levels than the first. In addition to this, achieving high levels of transduction in the CNS can be challenging. In this chapter we describe the cloning of a bicistronic AAV vector that uses the foot-and-mouth disease virus 2A sequence to efficiently express two genes from a single promoter. Bicistronic expression of a therapeutic gene and a reporter gene is desirable so that the axons from transduced neurons can be tracked and, after CNS injury, the amount of axonal sprouting or regeneration quantified. We go on to describe how to perform a pyramidotomy model of CNS injury and the injection of AAV vectors into the sensorimotor cortex to provide efficient transduction and bicistronic gene expression in cortical neurons such that transduced axons are detectable in the dorsal columns of the spinal cord.

Key words AAV, CNS injury, Regeneration-associated gene, Bicistronic vector, 2A sequence, Ribosomal skipping, Self-cleavage, Intracortical injection, Pyramidotomy

1 Introduction

Following an injury to the central nervous system (CNS) neurons undergo collateral sprouting but very limited long-distance regeneration which results in their failure to form functional connections with their original targets. This is due in part to the reduced intrinsic growth state of mature CNS neurons, which is characterized by their failure to express regeneration-associated genes (RAGs) after an injury [1–5]. Steady advancements in molecular neurobiology

Andrew J. Murray (ed.), *Axon Growth and Regeneration: Methods and Protocols*, Methods in Molecular Biology, vol. 1162,
DOI 10.1007/978-1-4939-0777-9_16,

have elevated gene therapy into a promising therapeutic strategy for increasing the expression of RAGs and improving regeneration after a CNS injury [6–9].

To test a potential gene therapy an in vivo model of CNS injury is required. The corticospinal tract (CST) is present in all mammals and originates from the pyramidal corticospinal neurons (CSNs) in layer V of the cerebral cortex that project axons through the brainstem to the spinal cord. The CST fibers descend through the ventral brainstem in the medullary pyramids after which the main component of the CST decussates at the spinomedullary junction to form the dorsal CST which runs in the ventral portion of the contralateral dorsal funiculus in the adult rat [10]. The CST is a descending motor pathway that primarily controls locomotion, posture, and voluntary skilled movements [11]. Following injury to the rodent CST, persistent deficits in sensorimotor function are observed and restorative strategies are often assessed using behavioural tests of sensorimotor and skilled motor function [11–14]. The CST can be lesioned by performing a unilateral pyramidotomy (i.e., one of the medullary pyramids is cut) [15]. This leads to the anterograde degeneration of the injured axons and functional deficits on the contralateral side of the body. These deficits can be overcome by enhancing the growth state of the uninjured CSNs, allowing them to sprout their axons across the midline and re-innervate the area denervated by the lesion. The pyramidotomy model has been extensively used to assess potential therapies to enhance neuronal regeneration and sprouting and functional recovery following CNS injury [8, 16]. However, efficient, widespread transduction of the CSNs in the cortex is challenging and has only been achieved when high titers (1×10^{14} genome copies (GC)/mL) and slow infusion rates were used [6, 8, 17]. Here we outline a procedure to inject AAV vectors into the cortex to transduce CSNs and then perform a pyramidotomy injury.

Recombinant adeno-associated viral (AAV) vectors represent one of the most attractive gene delivery systems for the CNS due to their ability to efficiently transduce neurons and to provide long-term expression with only a minimal immune response [18, 19]. AAV vectors have one of the best characterized safety profiles and have been taken to clinical trial for a number of neurodegenerative diseases [20, 21]. In addition to this, Glybera (an AAV vector encoding lipoprotein lipase) has recently been approved for use in the EU to reduce the incidence of pancreatitis in people with inherited lipoprotein lipase deficiency [22] and paves the way for further AAV-based gene therapies for other conditions. Each AAV serotype expresses different capsid proteins that determine which receptors the AAV vector can bind to for cell entry and therefore establishes the AAV vector's tropism [23]. Pseudotyping AAV2 genome with the capsid from different AAV serotypes can generate vectors with different tissue and cellular tropisms, which can improve the efficiency and pattern of transduction in specific regions of the CNS [24–26]. We have previously shown that AAV1 is the optimal serotype for transducing CSNs in adult rats [17] although other AAVs also work well: AAV5 [27–29] and AAV8 [6, 28, 30]. Viral vector-mediated over-expression of pro-regenerative genes, including retinoic acid receptor, beta (RARbeta2), neuronal calcium sensor 1 (NCS1), and Kruppel-like factor 7 (KLF-7), have been shown to enhance the intrinsic growth state of CSNs and facilitate axonal regeneration after injury [6–9]. However, these studies have either used viral vectors with larger insert capacities such as lentiviral vectors or used two separate AAV vectors: one encoding a RAG and the other a reporter gene. It would therefore be very valuable to have an AAV vectors which can co-express a pro-regenerative gene together with a fluorescent reporter so that transduced axons can be distinguished from axons that are not transduced: this would increase our ability to identify useful pro-regenerative therapies.

A disadvantage of AAV vectors is their relatively low insert capacity which can be an issue if two genes are required to be expressed (e.g., from two promoters). One partial solution is to use a single promoter which drives two genes linked by an internal ribosomal entry site (IRES) [31, 32]: typically, the RAG would be expressed immediately downstream of the promoter, whilst the fluorescent reporter gene is expressed downstream of the IRES. Although this works reasonably well, the IRES-dependent second gene is usually expressed at a significantly lower level, estimated at tenfold less than the cap-dependent first gene [33, 34]. To date we know of no reports where transduction of supraspinal neuronal cell bodies with such an AAV vector results in detectable labelling of their axons in the spinal cord.

This problem can be mitigated by using the 2A peptide from the foot-and-mouth disease virus (FMDV) [35]. The 2A peptide sequence is only 24 amino acids long and can mediate efficient bicistronic expression of two gene products from a single promoter by undergoing ribosomal skipping (referred to as "self-cleavage") during protein translation. Theoretically, the two gene products are expressed in a 1:1 ratio, and in practice each gene product is expressed at a high level. The 2A and 2A-like bicistronic systems have been shown to be highly effective in the CNS where high expression levels of both genes have been demonstrated [6, 34].

2 Materials

2.1 Chemicals and Solutions

Plasmids: The psubCMV-2A-WPRE plasmid (Fig. 1a) was a kind gift from Dr. Hansruedi Bueler, University of Kentucky. The plasmid contains a *cytomegalovirus* (CMV) promoter (to drive expression of the bicistron), the FMDV 2A sequence, and a woodchuck hepatitis virus post-transcriptional regulatory element (WPRE). The 2A sequence is immediately flanked by an upstream (i.e., 5′)

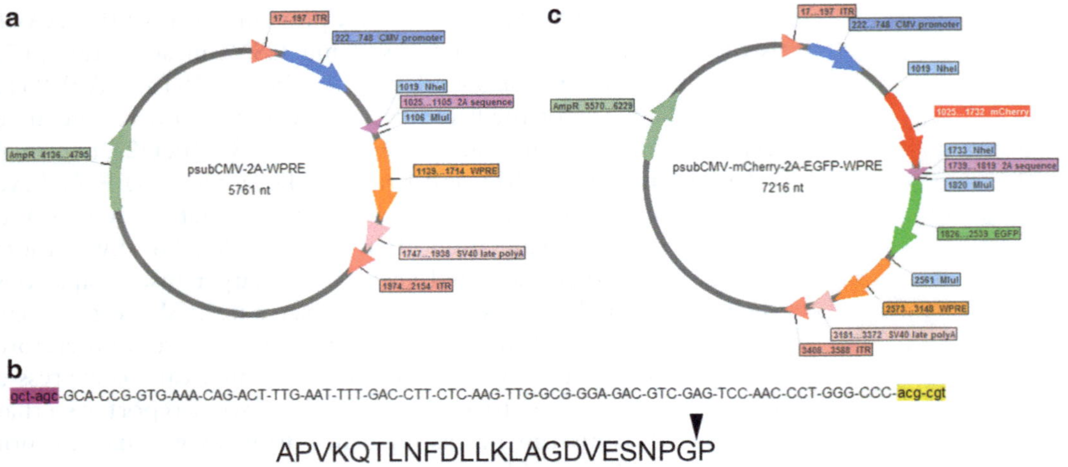

Fig. 1 (**a**) psubCMV-2A-WPRE plasmid map showing NheI site for cloning a transgene immediately downstream of the CMV promoter (and upstream of the 2A sequence) and an MluI site for cloning a transgene downstream of the 2A sequence. (**b**) Nucleotide sequence showing NheI restriction site (*lower case*, *pink*), the 2A sequence (*upper case*), and MluI restriction site (*lower case*, *yellow*). The 2A sequence encodes 24 amino acids APVKQTLNFDLLKLAGDVESNPGP, the *arrowhead* indicates the ribosome skipping ("self-cleavage") site. (**c**) psubCMV-mCherry-2A-EGFP-WPRE plasmid map showing mCherry cloned upstream and EGFP downstream of the 2A sequence

NheI restriction site and a downstream (i.e., 3′) MluI restriction site (Fig. 1b), *see* **Note 1**.

Enzymes: NheI and MluI, NEBuffer 4, and BSA (NEB, UK). Alkaline phosphatase (calf intestinal) (NEB, UK).

PCR reagents: Vent DNA polymerase (NEB, UK), 10× Vent DNA polymerase buffer, DMSO, 10 mM Nucleotide mix (Promega, UK), 100 mM $MgSO_4$, 100 μM forward and reverse primers, 50 ng plasmids containing mCherry or EGFP, PCR-grade nucleotide-free H_2O.

Anaesthetic: 0.5 mg/kg Domitor (medetomidine hydrochloride) and 100 mg/kg Vetalar (ketamine hydrochloride) or with a mixture of isoflurane (1.5 %) and O_2 (1 L/min).

Analgesia: 5 mg/kg Carprofen was administered subcutaneously for postoperative analgesia.

Euthanasia solutions: 400 mg/kg Lethabarb (sodium pentobarbital).

2.2 Cloning Strategy and Primer Design

Our goal was to generate a series of plasmids with the generic structure (AAV-RAG-2A-EGFP). First, we cloned EGFP downstream of 2A, and then we cloned mCherry upstream of 2A. The resulting plasmid (AAV-mCherry-2A-EGFP-WPRE) (Fig. 1c) has been used in many of our experiments as a negative control

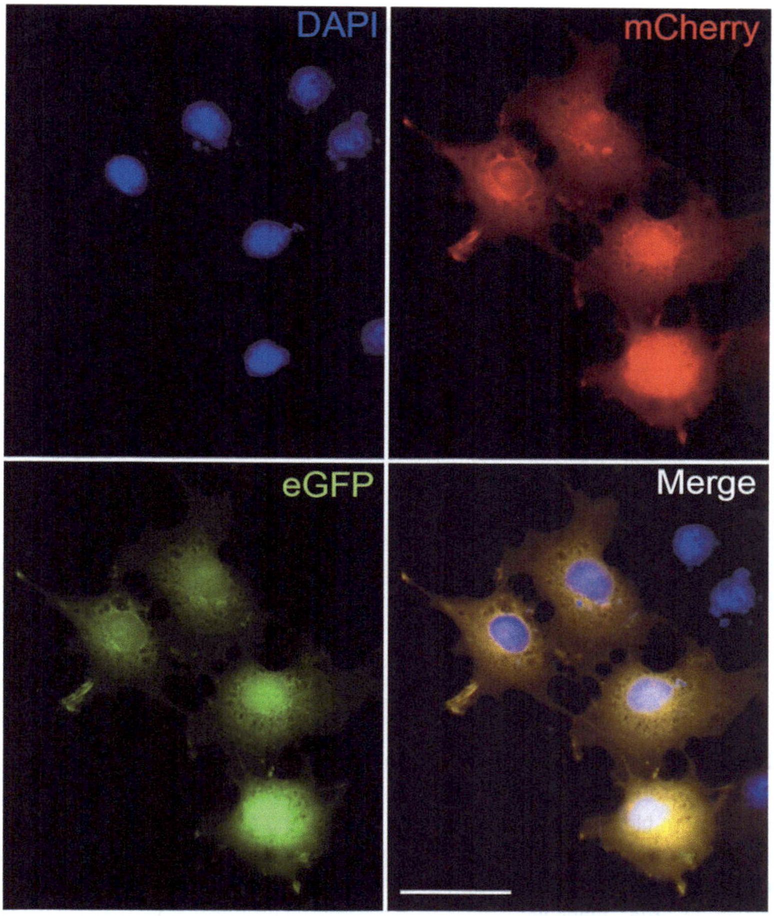

Fig. 2 COS-7 cells transfected with the psub-CMV-mCherry-2A-EGFP-WPRE plasmid efficiently express both mCherry and EGFP. Fluorescent images demonstrate that transfected cells express both mCherry and EGFP and appear *yellow*. Non-transfected cells do not express mCherry or EGFP but can be identified by DAPI staining. Scale bar—25 μm

(because mCherry is in place of a given RAG). It also enabled us to check easily that, indeed, both genes are expressed (Figs. 2 and 3), *see* **Note 2**. Subsequently, we have cloned more than ten different pro-regenerative transgenes in place of mCherry.

Forward and reverse primers were designed for EGFP (Fig. 4) and mCherry (Fig. 5) to ensure that the coding region of the plasmid produced one single open reading frame encoding all of the following: a start codon, the upstream cDNA (mCherry in this case), the 2A sequence, and the downstream cDNA (EGFP in this case), and a stop codon.

For cloning EGFP downstream of the 2A sequence, the forward primer was designed to contain a hexamer of GC repeats (blue) for stability of the PCR product and efficient restriction enzyme cleavage, an MluI restriction site (yellow), and the first

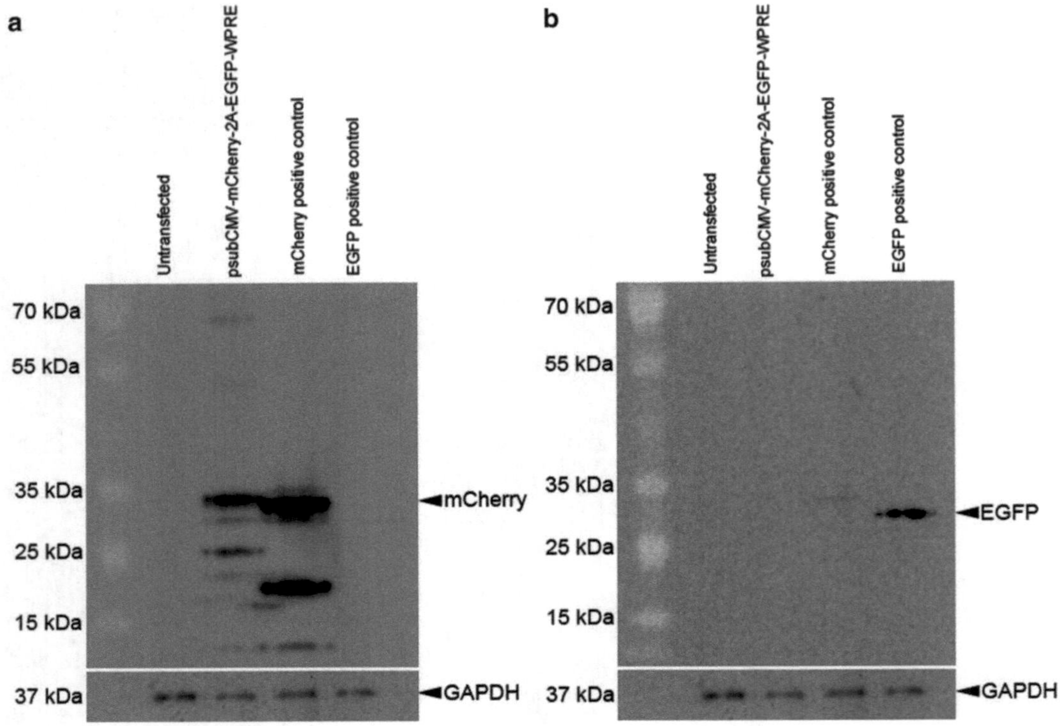

Fig. 3 Western blotting confirms efficient "self-cleavage" of the 2A sequence. COS-7 cells were transfected with the psub-CMV-mCherry-2A-EGFP-WPRE plasmid or the mCherry- or the EGFP-monocistronic plasmids. (**a**) Western blotting of 10 μg of protein lysate using an anti-mCherry antibody (Abcam—ab167453) detected a band at the expected size for mCherry protein (~30 kDa) and was a similar size as our mCherry-positive control (mCherry expressed from a monocistronic plasmid). The mCherry protein from the 2A plasmid was at a slightly larger molecular weight than the control mCherry protein which was expected due to the addition of 23 amino acids at its C-terminus following "self-cleavage" of the 2A sequence. Only a very faint band was detected at ~60 kDa, as would occur if an mCherry-2A-EGFP fusion protein was produced, demonstrating efficient 2A "self-cleavage" of the two proteins. Several unidentified bands were detected at smaller molecular weights for both the 2A plasmid and the mCherry-positive control. (**b**) We were unable to detect EGFP protein expression from the 2A plasmid via western blotting using several different anti-EGFP antibodies (Lifespan Bioscience—C183899, Clontech—632569, Abcam—ab13970, Sigma—G1546). However we did detect a band at the correct size for EGFP (~30 kDa) in our EGFP-positive control (EGFP expressed from a monocistronic plasmid)

28 bases of the EGFP cDNA excluding the ATG start codon, *see* **Note 3**. The reverse primer contained a hexamer of GC repeats (blue), an MluI restriction site (yellow), and the reverse complement of the last 28 bases of the EGFP cDNA (green) including the stop codon (bold and underlined) to ensure that translation ends after the synthesis of the bicistron.

For cloning mCherry upstream of the 2A, the forward primer was designed to contain a hexamer of GC repeats (blue) for stability of the PCR product and efficient restriction enzyme cleavage, a NheI restriction site (pink), and the first 28 bases of the mCherry

Forward:5′-GCG-CGC-ACG-CGT-GTG-AGC-AAG-GGC-GAG-GAG-CTG-TTC-ACC-G-3′

Reverse:5′-GCG-CGC-ACG-CGT-**<u>CTA</u>**-GTA-GGA-TCT-GAG-TCC-GGA-CTT-GTA-C-3′

Fig. 4 Primers used for cloning EGFP downstream of 2A

Forward:5′-GCG-CGC-GCT-AGC-**<u>ATG</u>**-GTG-AGC-AAG-GGC-GAG-GAG-GAT-AAC-A-3′

Reverse:5′-GCG-CGC-GCT-AGC-CTT-GTA-CAG-CTC-GTC-CAT-GCC-GCC-GGT-G-3′

Fig. 5 Primers used for cloning mCherry upstream of 2A

cDNA (red) including the start codon (ATG; bold and underlined). The reverse primer contained the GC repeat hexamer (blue), a NheI restriction site (pink), and the reverse complement of the last 28 bases of the mCherry cDNA with the stop codon excluded (red).

Primers were synthesized at 50 nmol with desalting but no other purification (Sigma, UK). We always check the resulting AAV plasmids by DNA sequencing to ensure that sequences are correct.

Sequencing primer used to check the success of cloning a gene upstream of 2A:

5′-AGC-TGC-GGA-ATT-GTA-CCC-GC-3′

Sequencing primer used to check the success of cloning a gene downstream of 2A:

5′-AAG-GCA-TTA-AAG-CAG-CGT-ATC-CAC-A-3′

2.3 Polymerase Chain Reaction Reagents

The reagents included in our standard PCR were as follows:

	Volume (μL)	Final concentration
DNA template	1	Determined by the user; we used 50 ng
Forward primer (100 μM)	1	1 μM
Reverse primer (100 μM)	1	1 μM
ThermoPol Reaction Buffer (10×)	10	1×
Vent polymerase (2 units/μL)	1	2 units
dNTPs (pool of 10 mM of each)	2	200 μM
DMSO	5	5 %
$MgSO_4$ (100 mM)	4	4 mM
Nuclease-free water	To a final volume of 100 μL	

2.4 Viral Vector Production and Titering

AAV vectors were generated by the Miami Project Viral Vector Core using fast protein liquid chromatography (FPLC)-based purification and pseudotyped with AAV capsid serotype 1, *see* **Note 4**. AAV vectors were suspended in HBSS at 2.9×10^{13} GC/mL, *see* **Note 5**.

2.5 Animals and Surgery

Animals: Female Lister hooded rats weighing 200–220 g (Charles River, UK) were used for these experiments. Rats were maintained under standard animal care conditions (12:12-h light/dark cycle), with food and water ad libitum. All procedures were carried out in accordance with the UK Animals (Scientific Procedures) Act 1986 and approved by the local veterinarian and ethical committee.

Brain atlas for stereotaxic coordinates: Holes were drilled through the skull above the sensorimotor cortex using coordinates reported in a microstimulation mapping study relative to bregma, midline, and the brain surface [36, 37] (*see* Subheading 3 for coordinates).

Stereotaxic apparatus: Stereotaxic frame for rats (World Precision Instruments, FL, USA) which allowed the attachment of a microdrill (Power Performance, UK) or ultra microsyringe pump with Micro4 control panel (World Precision Instruments, FL, USA).

Surgery materials: Eye lubricant (Viscotears, Germany), alcohol-based permanent marker, chlorhexidine disinfectant, sterile cotton buds, small bulldog clamps (World Precision Instruments # 14118), blunted scissors (Fine Science Tools, Fine Scissors—Tough Cut 14058-09), toothed forceps (Fine Science Tools, 11019-12), Long-toothed Alm retractors (Fine Science Tools, 17009-07), fine Dumont forceps (Fine Science Tools, 11251-10), dental hand drill, 0.7 mm drill bits (Fine Science Tools, 19007-07), Vannas Spring Scissors (Fine Science Tools, 15000-03), gelfoam (Equimedical, Gelatin Absorbable Haemostatic Sterile EQU705001), 3.0 Vicryl sutures (Ethicon), scalpel and #10 scalpel blades, needle holder, 27-G needles, 1 mL disposable syringes (for intraperitoneal (i.p.) injection of the anaesthetic), 10 μL Hamilton syringe (Hamilton, Reno, NV, USA), 15 mm, 34 gauge, point style 2, small hub removable needles (Hamilton, 207434), low-magnification stereomicroscope (2–20×).

2.6 Histology, and Microscopy

Fixative and buffers: 4 % paraformaldehyde in phosphate-buffered saline (PBS) pH 7 and cryoprotectant (30 % sucrose in PBS).

Embedding and sectioning: 10 % porcine gelatine in ddH_2O or cryoprotectant embedding medium OCT. A freezing microtome (Leitz, Wetzlar, Germany) or cryostat (Bright, UK) was used to cut the tissue.

Microscope: Confocal microscope (Carl Zeiss LSM 710).

3 Methods

3.1 Plasmid Design and Cloning

1. The coding sequences for mCherry and EGFP were amplified separately using primers and PCR reagents listed above at the cycling conditions recommended by the manufacturer for the Vent polymerase. The conditions for our typical PCR are shown below, *see* **Note 6**:

		Temperature (°C)	Time
1 cycle	Holding	95	2 min
25 cycles	Denaturation	95	30 s
	Annealing	54	30 s
	Extension	73	3 min
1 cycle	Final extension	73	11 min

2. An aliquot of the amplicons (2.5 μL of the 100 μL reaction) was analyzed using agarose gel electrophoresis to ensure that the PCR products were of the expected sizes.
3. These blunt-ended amplicons were separately cloned into the pCR-BluntII-TOPO plasmid (Invitrogen, UK) according to the manufacturer's instructions, creating pCR-BluntII-TOPO-mCherry and pCR-BluntII-TOPO-EGFP. Competent *E. coli* were then transformed using the heat-shock protocol and 3 μL of the cloning reaction. Transformants were streaked out onto LB-agar plates containing kanamycin (10 μg/mL).
4. A day later, putative positive clones were inoculated into 5 mL of LB-kanamycin and DNA was purified using standard minipreps (Promega, UK).
5. Clones were screened by restriction digests of the purified DNA using 0.5–1.0 μg of pCR-BluntII-TOPO-mCherry or pCR-BluntII-TOPO-EGFP digested by NheI or MluI, respectively. Our standard screening restriction digests were as follows, *see* **Note 7**:

Reagent	Volume
DNA (up to 1 μg)	1 μL
10× Buffer 4	2 μL
100× BSA	0.2 μL
Restriction enzyme (NheI or MluI)	0.5 μL (2–10 units)
Nuclease-free water	To a final volume of 20 μL

6. Clones were screened by performing an MluI or an NheI restriction digest and visualized using agarose gel electrophoresis to ensure that the cDNA inserts were present. Subsequently,

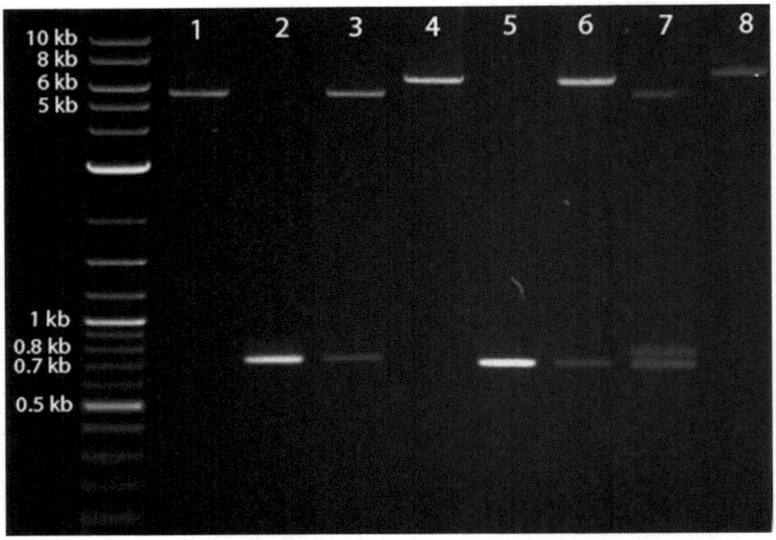

Fig. 6 Agarose gel electrophoresis of restriction-digested plasmid DNA depicts the expected products at each step when cloning mCherry and EGFP into the psub-CMV-2A-WPRE plasmid. See text for a detailed explanation of the steps illustrated in the figure

these clones were sequenced using the SP6 and T7 primer sites that flank the "multiple cloning site" of pCR-Blunt II-TOPO to ensure the fidelity of the PCR amplification.

7. At least 10 μg of the psub-CMV-2A-WPRE vector was digested with MluI and then treated with alkaline phosphatase to prevent re-circularization. Similarly, 10 μg of pCR-BluntII-TOPO-EGFP was also digested with MluI to release the EGFP cDNA. Both the vector (Fig. 6, lane 1) and the insert (Fig. 6, lane 2) were then purified using gel extraction.
8. The EGFP cDNA insert was then ligated into the vector using T4 DNA ligase, following the manufacturer's protocol, to create psub-CMV-2A-EGFP-WPRE.
9. This new plasmid was transformed into *E. coli*, and clones were screened for the presence of the insert using an MluI restriction digest (as described above) (Fig. 6, lane 3) and then sequenced to ensure the correct orientation of the EGFP insert.
10. Next, to clone in mCherry upstream of the 2A-EGFP sequence, psub-CMV-2A-EGFP-WPRE was digested with NheI and treated with alkaline phosphatase. The pCR-BluntII-TOPO-mCherry was also digested with NheI to release the mCherry cDNA.

11. The linearized psub-CMV-2A-EGFP-WPRE (Fig. 6, lane 4) and the mCherry cDNA insert (Fig. 6, lane 5) were gel purified and then ligated using T4 DNA ligase to create psub-CMV-mCherry-2A-EGFP-WPRE and transformed into *E. coli*. The following day, DNA minipreps were obtained for 10–20 colonies.
12. Clones were screened for the presence of the insert via NheI restriction digest (Fig. 6, lane 6) and then sequenced to ensure the correct orientation of mCherry.
13. psub-CMV-mCherry-2A-EGFP-WPRE was digested with MluI and NheI to show the presence of both EGFP and mCherry (Fig. 6, lane 7).
14. psub-CMV-mCherry-2A-EGFP-WPRE was digested with XhoI to illustrate the increased size of the linearized plasmid (Fig. 6, lane 8) relative to lanes 1 and 4.

3.2 Electroporation of Plasmids

1. Because it is expensive to generate AAV vectors, we routinely screen newly cloned AAV transfer plasmids in vitro using electroporation [38]. We can check that the plasmid is expressing the reporter protein (Fig. 2), and, if the plasmid contains a RAG, that it promotes neurite outgrowth, *see* **Note 8**.

3.3 Viral Vector Production and Titering

1. AAV vectors were generated by the Miami Project Viral Vector Core using FPLC-based purification.
2. The AAV vector transgene cassette was flanked by noncoding inverted terminal repeats (ITRs), which were the only wild-type viral sequences present. The AAV vectors were pseudotyped with AAV capsid serotype 1. Following production, the AAV vectors were titered using quantitative PCR to determine the number of genomic copies and an infection assay on 293 T cells was conducted to verify transduction efficiency.

3.4 Intracortical Injections

1. Female adult Lister hooded rats (Charles River, UK) weighing 200–220 g were anesthetized with an intraperitoneal injection of 0.5 mg/kg Domitor (medetomidine hydrochloride) and 100 mg/kg Vetalar (ketamine hydrochloride).
2. Once anesthetized, the head was shaved and sterilized using chlorhexidine disinfectant.
3. Eye lubricant was applied to the eyes, and throughout the surgery, the animals' body temperature was maintained at 37 °C with a thermal heating blanket connected to a rectal temperature probe.
4. The rats' head was placed into the stereotaxic frame, and the incisor bar was lowered 2 mm below the horizontal plane so

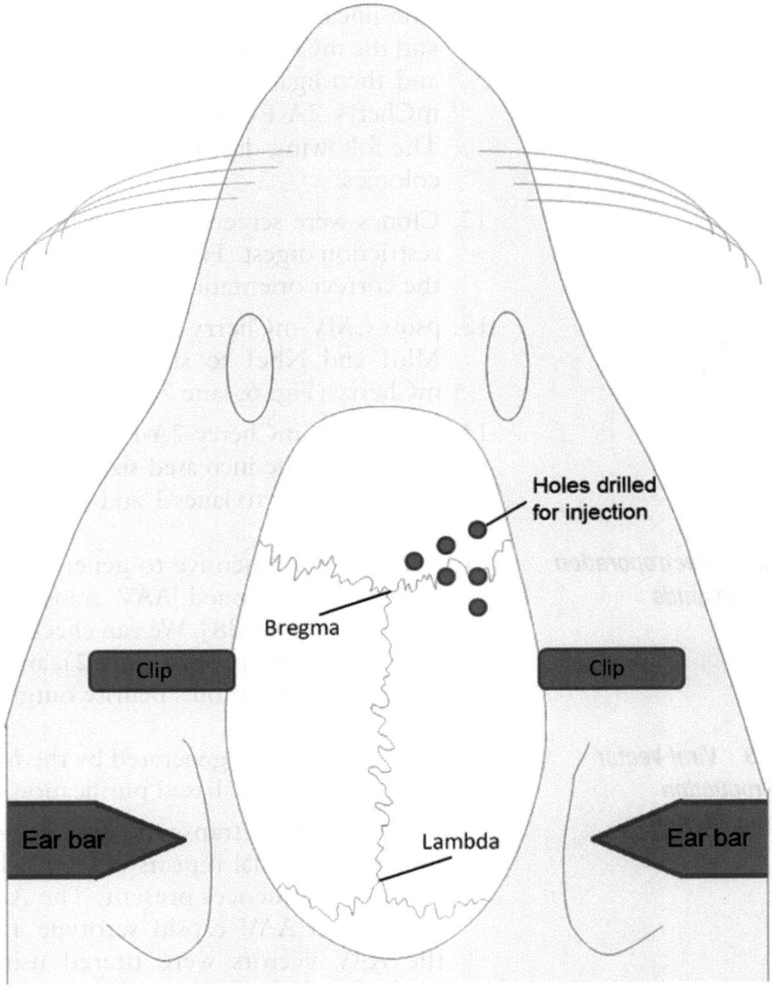

Fig. 7 Schematic showing the position of the holes drilled into the skull, relative to bregma, to expose the sensorimotor cortex for AAV vector injection. See text for a detailed explanation of the drilling and injection procedure

that the heights of lambda and bregma were equal and the surface of the skull was flat (Fig. 7).

5. Following a midline incision, the skin was retracted using small bulldog clamps and the periosteum was removed with a scalpel blade to expose the skull surface.
6. The microdrill was fixed to the stereotaxic frame holder and the drill burr centred over bregma. Six holes were then drilled through the skull above the left or the right sensorimotor cortex using coordinates reported in a microstimulation mapping study relative to bregma, midline, and the brain surface (Fig. 7). Defined as anterior–posterior (AP), medial–lateral (ML), and dorsal–ventral (DV):

First: 3.5 mm ML, −0.5 mm AP, DV −1.5 mm.

Second: 3.5 mm ML, +0.5 mm AP, DV −1.5 mm.

Third: 3.5 mm ML, +2 mm AP, DV −1.5 mm.

Fourth: 2.5 mm ML, +1.5 mm AP, DV −1.5 mm.

Fifth: 2.5 mm ML, +0.5 mm AP, DV −1.5 mm.

Sixth: 1.5 mm ML, +1 mm AP, DV −1.5 mm.

7. A 10 μL Hamilton syringe fitted with a 15 mm, 34 gauge, small hub removable needle was loaded with 3 μL of viral vector and fixed to the stereotaxic frame holder.
8. The tip of the needle was positioned over the first hole and lowered onto the brain surface. Once on the brain surface it was lowered a further 1.5 mm into the sensorimotor cortex.
9. 0.5 μL of AAV vector was infused into the brain using a microsyringe pump. The infusion rate was set to 200 nL/min on the Micro4 control panel, *see* **Note 9**.
10. Once the pump infusion was finished the needle was left in place for 3 min to allow the AAV vector to diffuse from the injection site.
11. The needle was then removed and positioned at the second coordinate to inject the next 0.5 μL of AAV vector. This process was repeated until all six injections were complete.
12. After the last injection the skin was closed using 3.0 vicryl sutures.
13. The rats were then placed into an incubator at 37 °C and 1 mg/kg Antisedan (atipamezole hydrochloride) was administered Subcutaneously. Once awake, 5 mg/kg Carprofen was administered subcutaneously for postoperative analgesia.

3.5 Pyramidotomy Injury Model

1. Rats were anesthetized using 5 % isoflurane in O_2 (flow rate 1.5 L/min) and maintained with 1–2 % isoflurane in O_2. Throughout the surgery the animals' body temperature was maintained at 37 °C with a thermal heating blanket connected to a rectal temperature probe, *see* **Note 10**.
2. The ventral neck region was shaved and sterilized using chlorhexidine disinfectant.
3. A midline incision is made from the chin to the rostral end of the sternum and the outer layers of gland, and muscle tissue is blunt dissected until the trachea is exposed.
4. The trachea is displaced slightly to one side, and the underlying muscle is dissected rostrally until the basioccipital skull is reached, *see* **Note 11**.
5. Surgical long-teethed retractors are inserted to allow access to the basioccipital bone.

6. A craniotomy of the ventral occipital bone was performed using a microdrill. This allows access to the medullary pyramids, which are located medially.
7. The left or the right pyramid was cut using Vannas microscissors making sure that the medial boundary close to the basilar artery is lesioned as well, *see* **Notes 12–14**.
8. Any bleeding is stopped, and the skin is sutured using 3.0 vicryl sutures.
9. The rats were placed in an incubator at 37 °C until fully awake. 5 mg/kg Carprofen was administered subcutaneously for postoperative analgesia.

3.6 Histology

1. 10 weeks post-injection, the rats were terminally anesthetized by intraperitoneal injection of 400 mg/kg Euthasol (sodium pentobarbital) and transcardially perfused with 200 mL PBS (pH 7.4) to remove the blood, followed by 200 mL 4 % paraformaldehyde in PBS (pH 7.4).
2. Immediately following perfusion the brain and spinal cord were dissected and post-fixed in 4 % paraformaldehyde in PBS for 2 h at 4 °C.
3. The tissue was then cryoprotected in 30 % sucrose with 0.1 % sodium azide for 5 days at 4 °C.
4. The brain and spinal cord tissue were placed in small plastic boxes and blocked in porcine gelatin.
5. Once the gelatin had set, they were placed overnight in 4 % paraformaldehyde in PBS at 4 °C to harden.
6. To allow tissue from each brain to undergo different immunohistological analysis the brains were cut rostro-caudally into 50 μm coronal serial sections using a freezing microtome and collected into ten series in a 24-well plate containing PBS and 0.1 % sodium azide. The spinal cords were cut rostro-caudally into 30 μm transverse serial sections and stored in the same manner.
7. The tissue sections were then mounted onto microscope slides using Mowiol and kept in the dark.

3.7 Microscopy

1. Slides were imaged using a confocal microscope (Fig. 8).

4 Notes

1. The cloning capacity of AAVs is approximately 5 kB from ITR to ITR, and longer sequences may lead to poor packaging and reduced titers. When using psubCMV-2A-EGFP-WPRE, only about 2.1 kb is available for a transgene. However, if longer

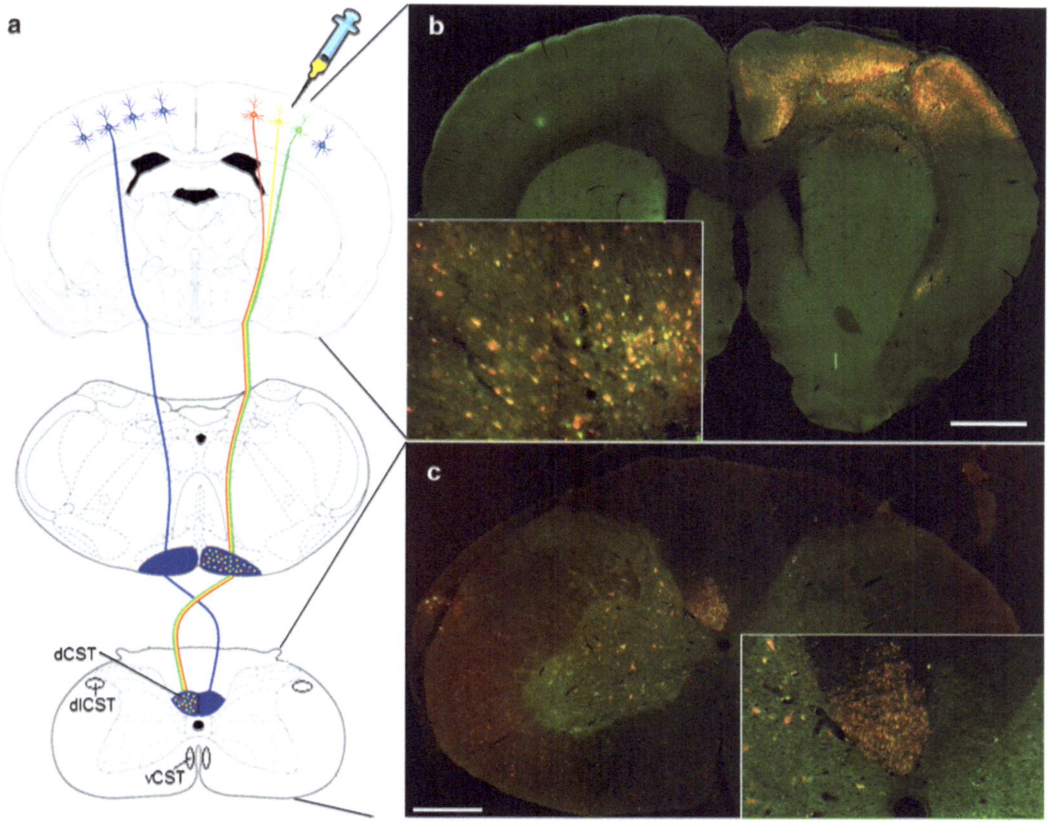

Fig. 8 (**a**) Schematic representation showing the rodent CST originating from the pyramidal CSNs in layer V of the sensorimotor cortex, decussating at the spinomedullary junction, forming the main dCST and the minor dlCST and vCST. (**b**) Image showing widespread transduction of the cortex; many cortical neurons including CSNs in layer V are expressing both mCherry and EGFP (cells appear *yellow*). (**c**) Image showing numerous mCherry- and EGFP-positive CST axons (axons appear *yellow*) in the contralateral dCST of the cervical spinal cord. Unexpectedly, some fibers in the dorsal columns appear to be just mCherry or EGFP positive. We are unsure why this is, but it may reflect differences in the vesicular transport of the two proteins. Others have reported similar issues (Wall et al [39] and references therein). Moreover, we have not yet been able to detect transgene positive axon collaterals in the spinal grey matter

transgenes need to be expressed, then WPRE could be deleted to make additional space.

2. EGFP protein could not be detected from the 2A plasmids via western blotting. This is a highly surprising result considering that the transfected COS-7 cells strongly expressed EGFP (Fig. 3) and our EGFP-positive control (EGFP expressed from a monocistronic plasmid) worked. Possible explanations for this may be because part of the "self-cleaved" 2A sequence remains attached to the N-terminus of EGFP and prevents binding to the anti-EGFP antibodies. However this is unlikely as we have tested several different antibodies (Lifespan Bioscience—C183899, Clontech—632569, Abcam—ab13970,

Sigma—G1546) raised against different regions of the EGFP protein including the N- and C-terminus, and the C-terminus of the downstream gene should not be affected by "self-cleavage" of the 2A sequence.

3. Note that our intermediate plasmid (psub-CMV-2A-EGFP) will not express EGFP in vitro or in vivo unless a transgene (containing an appropriate start codon) is cloned upstream of 2A.
4. It is important to choose the correct AAV serotype for the cell type you wish to transduce. We and others have previously shown that AAV1 and 5 are the optimal serotypes for transducing cortical neurons.
5. Using high-titer AAV vectors is a crucial factor for achieving high-level transduction, ideally 1×10^{13} GC/mL or higher.
6. The extension time of the PCR cycle was increased in instances where the cDNA was significantly larger. Some of the RAGs we cloned were over 3,000 bp long and required an increase in the extension time to 6 min.
7. The cDNA to be inserted must be checked to ensure that none of the restriction sites used for cloning (MluI and NheI) are present within the coding region. For example one of our RAGs contained an NheI site within its coding sequence. We replaced the NheI restriction site of the upstream and downstream primers with an SpeI restriction site that has an identical overhang to NheI. However, a ligation of DNA products digested by NheI and SpeI loses the restriction site for both enzymes and it cannot be subsequently restriction digested by either enzyme.
8. Note that "self-cleavage" of the 2A plasmid results in the upstream gene product having an additional 23 amino acids attached to the C-terminus and the downstream gene product having one additional amino acid attached to the N-terminus (Fig. 1b). The consequences of this additional sequence need to be verified empirically on a case-by-case basis, but in our experience, most pro-regenerative genes continue to show pro-regenerative effects in vitro after being shuttled from a monocistronic plasmid into the bicistronic psub-CMV-2A-EGFP-WPRE plasmid.
9. An important step in achieving high levels of transduction and a large area of transduction is the use of a micropump to slowly inject the viral vector over several minutes. In addition once the injection is complete it is also important to leave the needle in place for at least a couple of minutes to allow viral vector diffusion away from the injection sites.
10. The pyramidotomy surgery requires practice. High mortality rates are associated with the pyramidotomy surgery (*see* Steward et al. [40] and Kathe et al. [41]).
11. Displacement of the trachea can cause breathing difficulties or apnea. Giving Dopram (Dopram V-Drops, 20 mg/mL),

a respiratory stimulant, during the surgery can increase breathing activity. Furthermore, loosening of the retractors multiple times during the surgery can stabilize the animal.

12. When the pyramids are lesioned the basilar artery should be avoided. In addition to this, if the cut is made too deep it can affect the midbrain nuclei that are part of the breathing and cardiovascular centers.
13. Lesions are often incomplete. Repeating the cut with a 26 G needle including the fibers close to the basilar artery will prevent this to some extent.
14. Immunohistochemical staining for PKCγ can be used to verify completeness of the pyramidotomy lesion.

Acknowledgments

We would like to thank Dr. Vance Lemmon and the Miami Vector Core for production of the AAV vectors. We would also like to acknowledge support from a Research Councils UK Academic Fellowship (L.M.), the British Pharmacological Society's Integrative Pharmacology Fund (L.M), Friends of Guy's Hospital Research Grants (L.M.), and a grant from the Henry Smith Charity (L.M. and T.H.).

References

1. Chen DF, Jhaveri S, Schneider GE (1995) Intrinsic changes in developing retinal neurons result in regenerative failure of their axons. Proc Natl Acad Sci U S A 92:7287–7291
2. Nicholls JG, Saunders N (1996) Regeneration of immature mammalian spinal cord after injury. Trends Neurosci 19:229–234
3. Goldberg JL (2004) Intrinsic neuronal regulation of axon and dendrite growth. Curr Opin Neurobiol 14:551–557
4. Goldberg JL, Klassen MP, Hua Y et al (2002) Amacrine-signaled loss of intrinsic axon growth ability by retinal ganglion cells. Science 296: 1860–1864
5. Afshari FT, Kappagantula S, Fawcett JW (2009) Extrinsic and intrinsic factors controlling axonal regeneration after spinal cord injury. Expert Rev Mol Med 11:e37
6. Blackmore MG, Wang Z, Lerch JK et al (2012) Kruppel-like Factor 7 engineered for transcriptional activation promotes axon regeneration in the adult corticospinal tract. Proc Natl Acad Sci U S A 109:7517–7522
7. Yip PK, Wong LF, Pattinson D et al (2006) Lentiviral vector expressing retinoic acid receptor {beta}2 promotes recovery of function after corticospinal tract injury in the adult rat spinal cord. Hum Mol Genet 15:3107–3118
8. Yip PK, Wong LF, Sears TA et al (2010) Cortical overexpression of neuronal calcium sensor-1 induces functional plasticity in spinal cord following unilateral pyramidal tract injury in rat. PLoS Biol 8:e1000399
9. Moore DL, Blackmore MG, Hu Y et al (2009) KLF family members regulate intrinsic axon regeneration ability. Science 326:298–301
10. Tracey DJ (1995) Ascending and descending pathways in the spinal cord. The rat nervous system, 2nd edn. Academic, San Diego, CA, pp 67–77
11. Lemon RN (2008) Descending pathways in motor control. Annu Rev Neurosci 31: 195–218
12. Whishaw IQ, Gorny B, Sarna J (1998) Paw and limb use in skilled and spontaneous reaching after pyramidal tract, red nucleus and combined lesions in the rat: behavioral and

anatomical dissociations. Behav Brain Res 93: 167–183
13. Whishaw IQ, Metz GA (2002) Absence of impairments or recovery mediated by the uncrossed pyramidal tract in the rat versus enduring deficits produced by the crossed pyramidal tract. Behav Brain Res 134: 323–336
14. Starkey ML, Barritt AW, Yip PK et al (2005) Assessing behavioural function following a pyramidotomy lesion of the corticospinal tract in adult mice. Exp Neurol 195:524–539
15. Thallmair M, Metz GAS, Z'Graggen WJ et al (1998) Neurite growth inhibitors restrict plasticity and functional recovery following corticospinal tract lesions. Nat Neurosci 1: 124–131
16. Cafferty WB, Strittmatter SM (2006) The Nogo-Nogo receptor pathway limits a spectrum of adult CNS axonal growth. J Neurosci 26:12242–12250
17. Hutson TH, Verhaagen J, Yanez-Munoz RJ et al (2012) Corticospinal tract transduction: a comparison of seven adeno-associated viral vector serotypes and a non-integrating lentiviral vector. Gene Ther 19:49–60
18. McCown TJ (2005) Adeno-associated virus (AAV) vectors in the CNS. Curr Gene Ther 5:333–338
19. Papale A, Cerovic M, Brambilla R (2009) Viral vector approaches to modify gene expression in the brain. J Neurosci Methods 185:1–14
20. Kaplitt MG, Feigin A, Tang C et al (2007) Safety and tolerability of gene therapy with an adeno-associated virus (AAV) borne GAD gene for Parkinson's disease: an open label, phase I trial. Lancet 369:2097–2105
21. Lim ST, Airavaara M, Harvey BK (2009) Viral vectors for neurotrophic factor delivery: a gene therapy approach for neurodegenerative diseases of the CNS. Pharmacol Res 61:14–26
22. Yla-Herttuala S (2012) Endgame: glybera finally recommended for approval as the first gene therapy drug in the European union. Mol Ther 20:1831–1832
23. Vandenberghe LH, Wilson JM, Gao G (2009) Tailoring the AAV vector capsid for gene therapy. Gene Ther 16:311–319
24. Burger C, Nash K, Mandel RJ (2005) Recombinant adeno-associated viral vectors in the nervous system. Hum Gene Ther 16: 781–791
25. Cearley CN, Wolfe JH (2006) Transduction characteristics of adeno-associated virus vectors expressing cap serotypes 7, 8, 9, and Rh10 in the mouse brain. Mol Ther 13:528–537
26. Rabinowitz JE, Rolling F, Li C et al (2002) Cross-packaging of a single adeno-associated virus (AAV) type 2 vector genome into multiple AAV serotypes enables transduction with broad specificity. J Virol 76:791–801
27. Foust KD, Flotte TR, Reier PJ et al (2008) Recombinant adeno-associated virus-mediated global anterograde delivery of glial cell line-derived neurotrophic factor to the spinal cord: comparison of rubrospinal and corticospinal tracts in the rat. Hum Gene Ther 19:71–82
28. Blits B, Derks S, Twisk J et al (2010) Adeno-associated viral vector (AAV)-mediated gene transfer in the red nucleus of the adult rat brain: comparative analysis of the transduction properties of seven AAV serotypes and lentiviral vectors. J Neurosci Methods 185:257–263
29. Burger C, Gorbatyuk OS, Velardo MJ et al (2004) Recombinant AAV viral vectors pseudotyped with viral capsids from serotypes 1, 2, and 5 display differential efficiency and cell tropism after delivery to different regions of the central nervous system. Mol Ther 10:302–317
30. Taymans JM, Vandenberghe LH, Haute CV et al (2007) Comparative analysis of adeno-associated viral vector serotypes 1, 2, 5, 7, and 8 in mouse brain. Hum Gene Ther 18: 195–206
31. Klein RL, Lewis MH, Muzyczka N et al (1999) Prevention of 6-hydroxydopamine-induced rotational behavior by BDNF somatic gene transfer. Brain Res 847:314–320
32. Martinez-Salas E (1999) Internal ribosome entry site biology and its use in expression vectors. Curr Opin Biotechnol 10:458–464
33. Mizuguchi H, Xu Z, Ishii-Watabe A et al (2000) IRES-dependent second gene expression is significantly lower than cap-dependent first gene expression in a bicistronic vector. Mol Ther 1:376–382
34. Furler S, Paterna JC, Weibel M et al (2001) Recombinant AAV vectors containing the foot and mouth disease virus 2A sequence confer efficient bicistronic gene expression in cultured cells and rat substantia nigra neurons. Gene Ther 8:864–873
35. Minskaia E, Nicholson J, Ryan MD (2013) Optimisation of the foot-and-mouth disease virus 2A co-expression system for biomedical applications. BMC Biotechnol 13:67
36. Paxinos G, Watson C (1982) The rat brain in stereotaxic coordinates. Academic, San Diego, CA
37. Neafsey EJ, Bold EL, Haas G et al (1986) The organization of the rat motor cortex: a microstimulation mapping study. Brain Res 396: 77–96

38. Hutson TH, Buchser WJ, Bixby JL et al (2011) Optimization of a 96-well electroporation assay for postnatal rat CNS neurons suitable for cost-effective medium-throughput screening of genes that promote neurite outgrowth. Front Mol Neurosci 4:55
39. Wall NR, Wickersham IR, Cetin A et al (2010) Monosynaptic circuit tracing in vivo through Cre-dependent targeting and complementation of modified rabies virus. PNAS 107:21848–21853
40. Steward O, Sharp K, Yee KM (2012) A re-assessment of the effects of intracortical delivery of inosine on transmidline growth of corticospinal tract axons after unilateral lesions of the medullary pyramid. Exp Neurol 233: 662–673
41. Kathe C, Hutson TH, Chen Q, Shine HD, McMahon SB, Moon, LDF, Unilateral pyramidotomy of the corticospinal tract in rats for assessment of neuroplasticity inducing therapies. Accepted for publication in JoVe

Chapter 17

Application of Tissue Clearing and Light Sheet Fluorescence Microscopy to Assess Optic Nerve Regeneration in Unsectioned Tissues

Xueting Luo, Benjamin Yungher, and Kevin K. Park

Abstract

Optic nerve crush injury, as a model to study central nervous system (CNS) injury, is widely used to assess potential therapeutic strategies, aimed at promoting axon regeneration and neuronal survival. Traditional methods to evaluate optic nerve regeneration rely on histological sectioning. However, tissue sectioning results in inevitable loss of three-dimensional (3D) information, such as axonal trajectories and terminations. Here we describe a protocol for whole-tissue assessment of optic nerve regeneration in adult mice without the need for histological sectioning.

Key words Axon regeneration, Light sheet fluorescence microscopy, Optical clearing, Tissue clearing, Ultramicroscope, Retinal ganglion cells, PTEN

1 Introduction

Traumatic or neuropathic (e.g., glaucomatous) damage to optic nerves results in degeneration of retinal ganglion cell (RGC) axons and the subsequent loss of neuronal circuits. Accordingly, functional recovery after optic nerve injury requires the damaged axons to regenerate, to navigate properly through the degenerating tissues, and to reconnect with their appropriate targets. It is critical to analyze the pattern of regenerating axons when evaluating potential therapeutic strategies.

Traditional methods of pattern analysis of optic nerve regeneration depend upon tissue sectioning. However, this technique involves major technical limitations, including axon fragmentation and the unavoidable loss of 3D information. Thus, identifying and reconstructing nonlinear axon growth characteristics like axon turning and branching require examination of multiple tissue sections. Such efforts are laborious and often unreliable.

Andrew J. Murray (ed.), *Axon Growth and Regeneration: Methods and Protocols*, Methods in Molecular Biology, vol. 1162,
DOI 10.1007/978-1-4939-0777-9_17, © Springer Science+Business Media New York 2014

Overcoming these technical constraints while preserving spatial information requires optical clearing of tissues and high-resolution whole-mount imaging. Several methods for optical clearing have been developed. For example, the aquatic urea solution *Scale* has been employed to map neuronal connections in the whole brain [1]. *3DISCO* takes advantage of organic solvent tetrahydrofuran (THF) to dehydrate tissues and benzyl alcohol/benzyl benzoate (BABB) for optical clearing. This method allows clearing of mouse spinal cord within a week while preserving tissue integrity. It has been used to track regenerating axons in whole injured spinal cord and to image fluorescently labeled cells (neurons and glia) at high resolution [2, 3]. In place of BABB, dibenzyl ether (DBE) is also used as a clearing regent [4]. *CLARITY* is novel in its method of lipid removal via electrophoresis and provides hydrogel support for the subcellular structure. This procedure, which takes less than a week, has the major advantage of being compatible with traditional molecular biology methods such as immunohistochemistry and in situ hybridization. Moreover, the processed tissues remain in an aquatic environment that preserves fluorescent signals [5]. In addition, *ClearT*, which is composed of formamide, has been shown to effectively clear embryos, whole mounts, and thick brain sections with minimal volume changes [6].

Along with the technical improvement in tissue clearing, multiphoton microscopy and light sheet fluorescence microscopy (LSFM) further facilitate imaging of axons in whole tissues [7, 8]. With these technical advancements, we have applied an organic solvent-based clearing protocol to survey axon regeneration patterns after optic nerve injury in mice [9]. Here we describe a detailed protocol for axon labeling, whole tissue dehydration, clearing, and imaging with LSFM for the assessment of optic nerve regeneration in adult tissues.

2 Materials

2.1 Animals and Regeneration-Inducing Factors

Adult PTEN/SOCS3$^{f/f}$ mice [10] were used to conditionally delete two genes, phosphatase and tensin homologue (*pten*) and suppressor of cytokine signaling-3 (*socs3*) specifically in RGCs. Deletion of these genes using adeno-associated virus (AAV) expressing Cre recombinase (1–4 × 10^{13} genome copies/mL) combined with intravitreal injection of recombinant ciliary neurotrophic factor (CNTF; 1 μg/μL; Pepro Tech) induces robust regeneration of RGC axons after intraorbital optic nerve crush [10, 11].

2.2 Chemicals and Solutions

1. 1 % cholera toxin subunit β (CTB), Alexa Fluor 555 (or Alexa Fluor 488) conjugate (C-22843 or C-34775, Invitrogen): Dissolve CTB-555 (or CTB-488) in sterile 1× phosphate-buffered saline (PBS) to a final concentration of 1 % (wt/vol).

2. 4 % paraformaldehyde (PFA): To make 1 L, dissolve 40 g of PFA (P6148, Sigma) in 1× PBS, adjust pH to 7.4 with NaOH, and filter.
3. THF gradient: Immediately before use, mix THF solution (186562, Sigma) with distilled H_2O to obtain 50, 70, and 80 % (vol/vol) THF gradient (*see* **Notes 1–3**).
4. BABB: Immediately before use, mix benzyl alcohol (305197, Sigma) and benzyl benzoate (B6630, Sigma) at a 1:2 ratio (*see* **Notes 2** and **3**).

2.3 Equipment

1. 50 μL Hamilton syringes (Hamilton).
2. Glass micro-pipettes (Sutter Instrument).
3. Surgical micro-forceps and micro-scissors (Fine Science Tools).
4. Orbital shaker.
5. Amber glass bottle.
6. Ultramicroscope, also known as LSFM (LaVision BioTech, Germany).
7. Software: Imaris v7.6 (Bitplane).

3 Methods

All experimental procedures are performed in compliance with IACUC approval at the University of Miami Miller School of Medicine. For all surgical procedures, mice are anesthetized with ketamine (60 mg/kg) and xylazine (10 mg/kg). Animals are recovered on 37 °C heat pad after surgery and receive subcutaneous injection of buprenorphine (0.05 mg/kg) as a postoperative analgesic.

3.1 Animals and Surgery

1. PTEN/SOCS3$^{f/f}$ mice (5–6 weeks old) received an intravitreal injection of AAV serotype 2-Cre plus recombinant CNTF in the left eye using a micro-syringe and glass pipettes. Pipette is inserted in peripheral retina just behind the ora serrata and angled towards the mid-vitreous body to avoid damage to the lens and retina. AAV2 delivery into the vitreous chamber allows expression of Cre recombinase and deletion of floxed genes in RGCs [10, 12, 13].
2. 2 weeks after AAV injection, left optic nerve is exposed and crushed intraorbitally approximately 1 mm from the optic disk with micro-forceps for 10 s. Between 2 and 8 weeks after injury, animals are euthanized.
3. 2 days before euthanasia, approximately 2 μL of CTB-555 is injected intravitreally using a micro-syringe to anterogradely label regenerating RGC axons.

Table 1
Steps for dehydration and clearing of mouse CNS tissues

	Dehydration						Clearing
	THF concentrations						BABB
	50 %	70 %	80 %	100 %	100 %	100 %	
Brain	12 h	12 h	12 h	12 h	12 h	12 h	12 h
Optic nerve	20 min	20 min	20 min	20 min	20 min	–	20 min

3.2 Tissue Preparation

1. Mice are anesthetized and perfused transcardially with 20 mL of PBS, followed by 40 mL of 4 % PFA at 5 mL/min (*see* **Note 4**).
2. After perfusion, tissues (i.e., eyes, optic nerves, and brain) are dissected out (*see* **Notes 5** and **6**) and post-fixed in 4 % PFA overnight at 4 °C.
3. Tissues are rinsed with cold PBS three times and stored in cold PBS until use.

3.3 Tissue Dehydration and Clearing

1. Tissues undergo dehydration in increasing concentrations of THF in amber glass bottles. To achieve optimal dehydration, incubation time for each step varies depending on the size of the tissue. We follow procedures as outlined in Table 1 (*see* **Note 7**). Metal forceps are used to transfer tissues (*see* **Note 8**). Incubations are carried out on an orbital shaker at room temperature (*see* **Note 9**).
2. Tissues are rendered relatively transparent at the end of BABB incubation (Fig. 1a–c). Transfer the tissue into fresh BABB before proceeding to imaging.

3.4 LSFM Imaging

1. Mouse brain and attached optic nerves are mounted onto the sample holder (LaVision) and immobilized with a screw (Fig. 1d–g). Cleared brain is placed with the ventral side on the top to allow imaging with a top-mounted camera (*see* **Note 10**).
2. Sample is loaded into a bath chamber filled with BABB (Fig. 1e).
3. Ultramicroscope generates a thin light sheet to illuminate biological samples from the side while imaging the excited plane by a microscope perpendicular to the light sheet. Moving the sample through the light sheet generates a 3D image stack. While the laser sheet scans through the whole tissue in *Z*-axis, multiple optical sliced images are collected (Figs. 2b–g and 3c–j). Images are taken in a 3 μm *Z*-increment. Samples are imaged with Coherent Sapphire laser 488 and 561 or filters 525/50 M and 620/60 M.
4. If multichannel imaging is desired, cleared tissue can be scanned sequentially by lasers of different wavelengths.

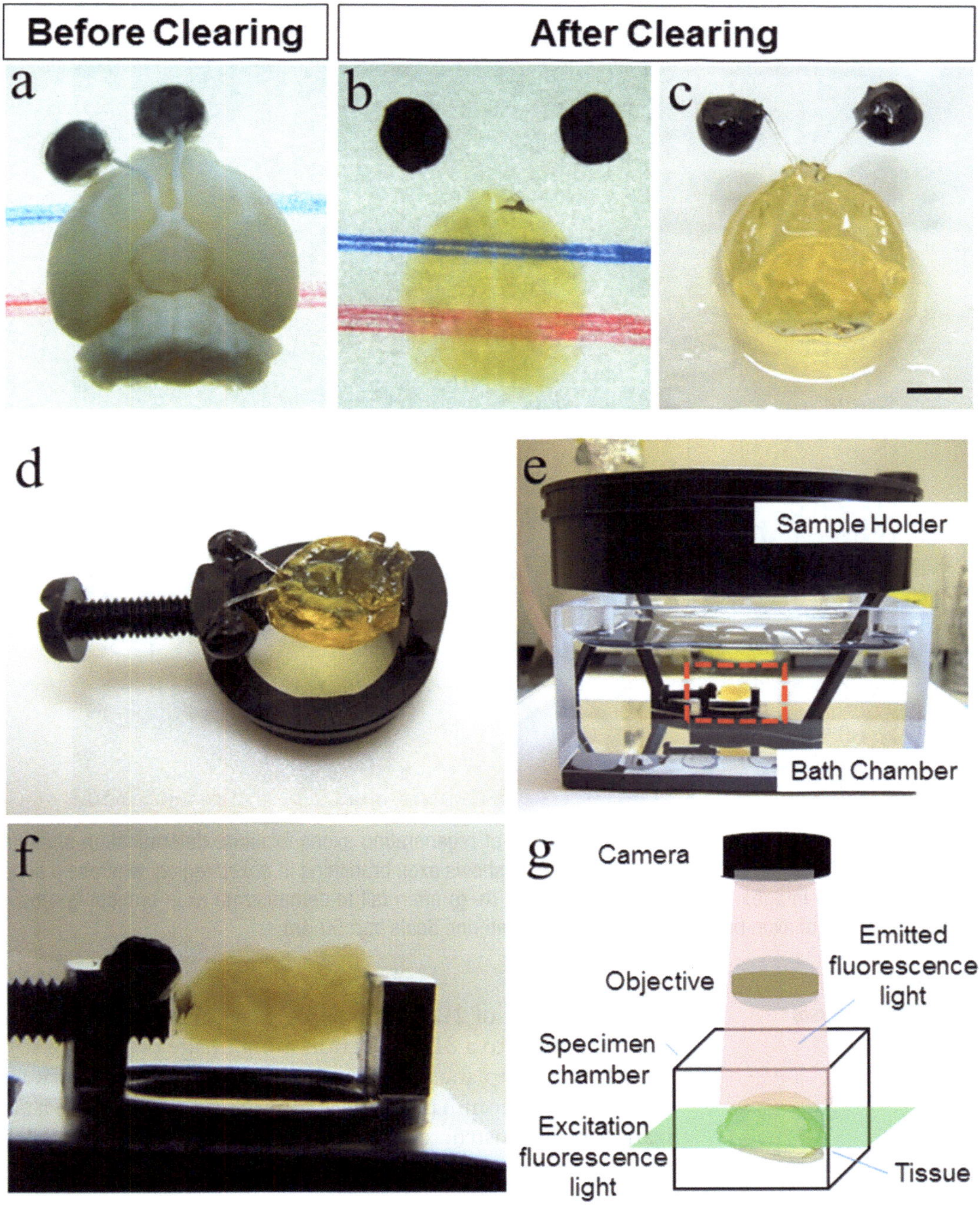

Fig. 1 Whole-tissue imaging using light sheet fluorescence microscopy (LSFM). (**a–c**), Mouse brain and optic nerves before and after tissue clearing. (**b**) Tissue in BABB. (**c**) Tissue in open air. Cleared tissue is firmly positioned with screw (**d**) and loaded into the sample holder and bath chamber (**e**, **f**). (**g**) Schematic representation of ultramicroscopy setup showing tissue positioning and the light path. Scale bar: 5 mm

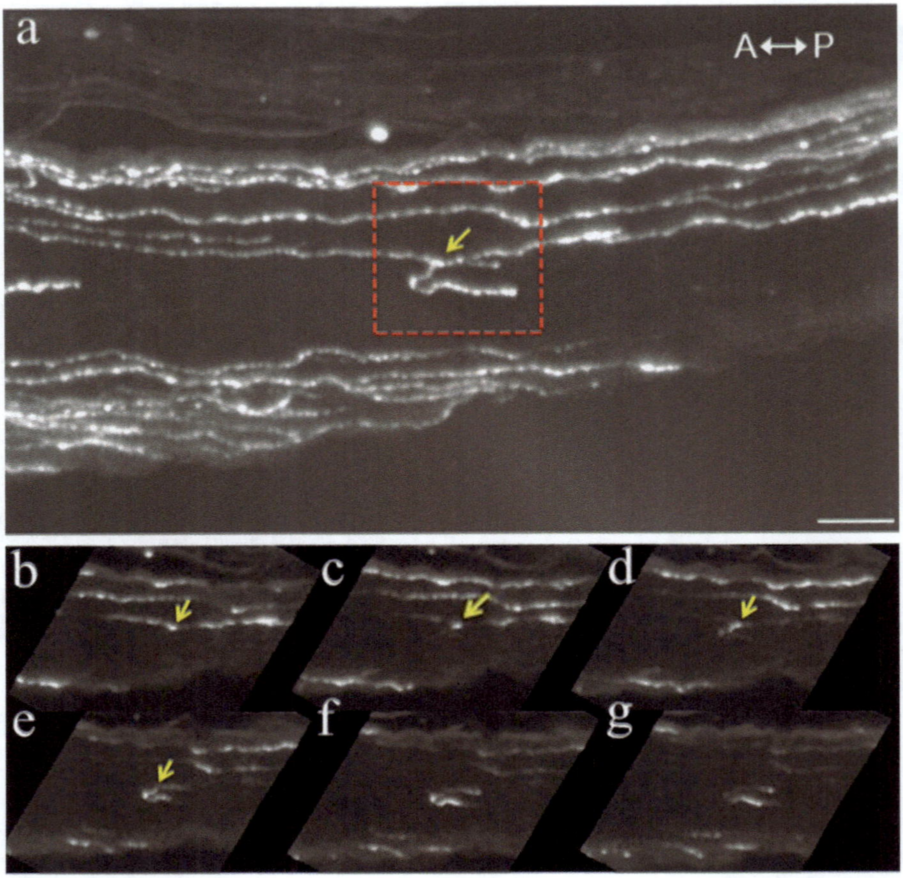

Fig. 2 Whole-tissue imaging and 3D reconstruction of regenerating axons facilitate determination of axonal branching. (**a**) 3D visualization of injured optic nerve shows axon branching in *boxed* region, whereas a series of consecutive 2D images taken at 3 μm increment (**b–g**) often fail to demonstrate axon branching. *Arrows* indicate the point of axon branching. *A* anterior, *P* posterior. Scale bar: 50 μm

3.5 Data Analysis

1. Reconstruction of 2D optical slices: Stacks of 2D optical slices are compiled into a 3D projection using image processing software Imaris (Bitplane) (*see* **Notes 11** and **12**). In this software, the "Surpass" command allows volume rendering and visualization of reconstructed whole tissue in 3D (Figs. 2a, 3a, b, and 4c).
2. Axonal tracing: Individual CTB-labeled axons are visible after 3D reconstruction. Axons can also be highlighted by tracing with "Filament Tracer" tool (Imaris v7.6). "Semiautomatic" tracing method tracks axon signals through the individual 2D optical slices (Fig. 3c–j).
3. Determination of axonal trajectories: Section-based histological methods to analyze regenerating axons inevitably result in the loss of spatial information such as trajectories. On a typical tissue section, axons are fragmented and scattered,

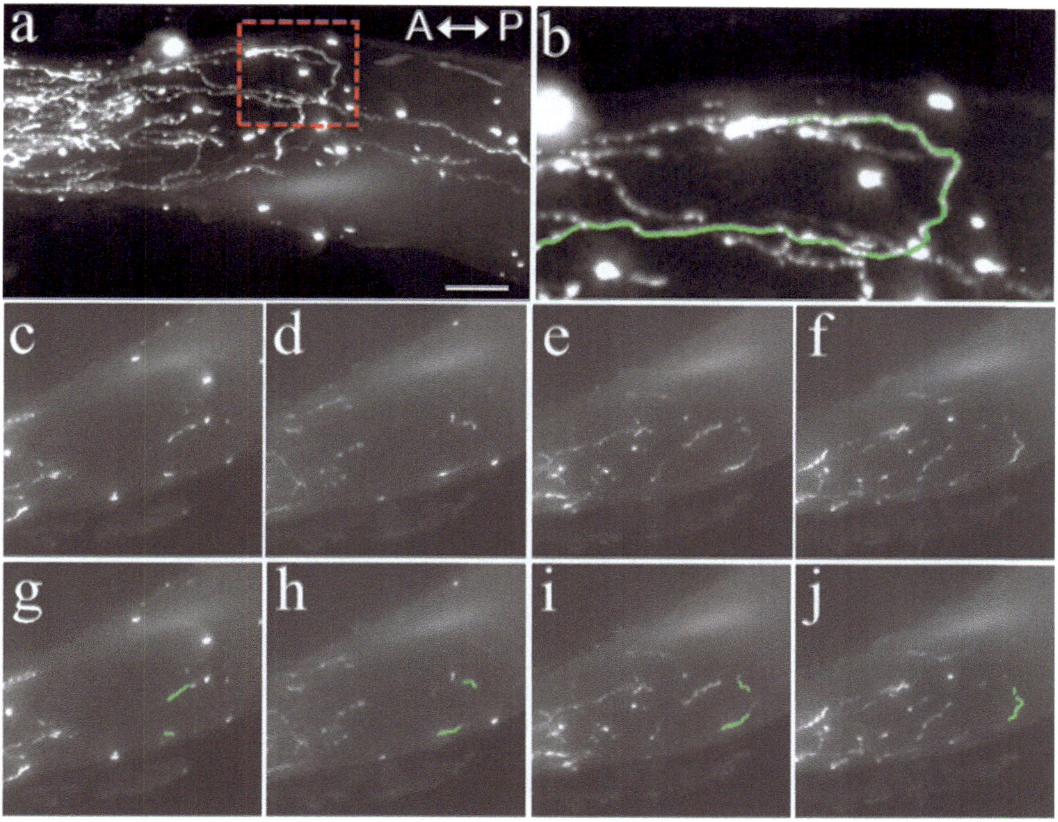

Fig. 3 Whole-tissue imaging and 3D reconstruction of regenerating axons facilitate determination of axon turnings. (**a**, **b**) 3D visualization of injured optic nerve shows RGC-elongating axons making multiple turns. An axon making a U-turn is highlighted in *red boxed* region (**b**). A series of consecutive 2D images taken at 3 μm increment (**c**–**j**) only show axon fragments as highlighted in *green tracing*. *A* anterior, *P* posterior. Scale bar: 50 μm

making axon navigation and targeting difficult to recognize. For example, while 3D visualization clearly shows axons making branches or making U-turns (Figs. 2a and 3a), 2D images from which the 3D images are reconstructed show lack of apparent signs of axon branching or U-turns (Figs. 2b–g and 3c–j). To reestablish neural circuit, regenerating axons must reach their targets. In Fig. 4, regenerating RGC axons are found around and within the suprachiasmatic nuclei (SCN, dotted area), a central target that receives photic input from RGCs. In 3D visualization (Fig. 4c), the locations of axon terminals can be determined precisely.

4 Notes

1. THF forms potentially explosive peroxides over time. Do not store THF stock or waste solutions for extended periods of time.
2. Use glass pipettes for transferring THF and BABB solutions.

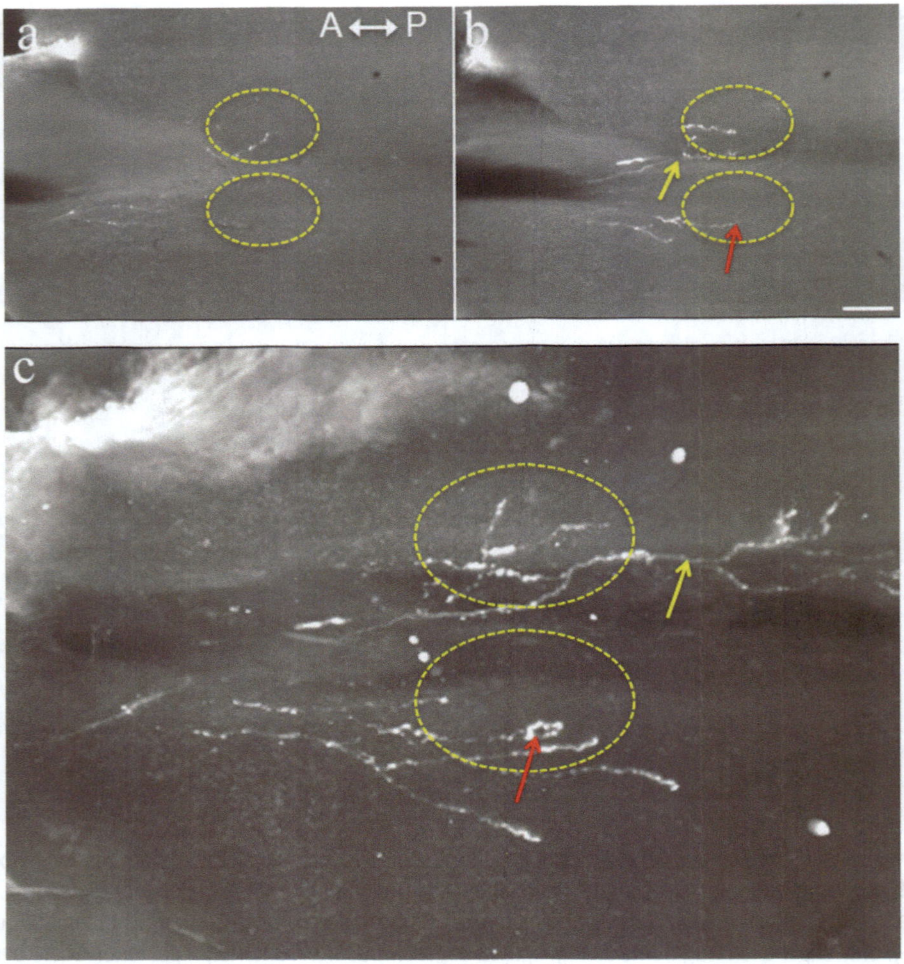

Fig. 4 Whole-tissue imaging and 3D reconstruction of regenerating axons facilitate determination of axon terminals. (**a**, **b**) 2D images show fragments of regenerating RGC axons, as indicated by *yellow* and *red arrows*, reaching the suprachiasmatic nucleus (SCN, *yellow circles*). In 3D visualization (**c**), some regenerating RGC axons navigate through the SCN (*yellow arrows*), while other axons terminate at the SCN (*red arrows*), possibly reforming synapses. *A* anterior, *P* posterior. Scale bar: 50 μm

3. THF and BABB solutions should be made fresh immediately before use to avoid peroxide degradation of fluorescent signal.
4. Perfusion with 4 % PFA should proceed at low speed. This allows optimal protein cross-linking and tissue fixation, which is beneficial to tissue integrity during subsequent manipulations.
5. Cautions should be exercised during dissections because tissues are vulnerable to mechanical damage, such as stretching and bending. Tissue damage results in artificial holes and axon fragmentation.

6. Remove the meninges as thoroughly as possible, as they are a major source of auto-fluorescence that compromises signal visibility after 3D reconstruction.
7. Avoid extended delays in the procedure. Extended storage of processing tissue in THF gradient or BABB may result in loss of signal intensity, tissue transparency, and image resolution.
8. As dehydration proceeds, tissue samples become fragile and must be handled with care.
9. Protect samples from light.
10. During laser scanning, samples should be immobilized firmly enough to avoid vibration without being compressed.
11. Data sets often comprise hundreds of 2D images and should be analyzed on a computer with sufficient RAM. Moreover, an advanced graphics card and high-resolution monitor would facilitate image recording.
12. Alternative commercially available software may be suitable for data analysis, such as Amira (Visage Imaging) and Volocity (PerkinElmer).

Acknowledgments

This work was supported by grants from the US Army W81XWH-12-1-0319 (KKP), NEI (KKP), and Ziegler Foundation (KKP). We thank the Imaging and Viral Vector Core at the Miami Project to Cure Paralysis.

References

1. Hama H, Kurokawa H, Kawano H, Ando R, Shimogori T, Noda H, Fukami K, Sakaue-Sawano A, Miyawaki A (2011) Nat Neurosci 14:1481–1488
2. Erturk A, Mauch CP, Hellal F, Forstner F, Keck T, Becker K, Jahrling N, Steffens H, Richter M, Hubener M, Kramer E, Kirchhoff F, Dodt HU, Bradke F (2012) Nat Med 18: 166–171
3. Erturk A, Bradke F (2013) Exp Neurol 242: 57–64
4. Becker K, Jahrling N, Saghafi S, Weiler R, Dodt HU (2012) PLoS One 7:e33916
5. Chung K, Wallace J, Kim SY, Kalyanasundaram S, Andalman AS, Davidson TJ, Mirzabekov JJ, Zalocusky KA, Mattis J, Denisin AK, Pak S, Bernstein H, Ramakrishnan C, Grosenick L, Gradinaru V, Deisseroth K (2013) Nature 497: 332–337
6. Kuwajima T, Sitko AA, Bhansali P, Jurgens C, Guido W, Mason C (2013) Development 140: 1364–1368
7. Becker K, Jahrling N, Kramer ER, Schnorrer F, Dodt HU (2008) J Biophotonics 1:36–42
8. Santi PA (2011) J Histochem Cytochem 59: 129–138
9. Luo X, Salgueiro Y, Beckerman SR, Lemmon VP, Tsoulfas P, Park KK (2013) Exp Neurol 247:653–662
10. Sun F, Park KK, Belin S, Wang D, Lu T, Chen G, Zhang K, Yeung C, Feng G, Yankner BA, He Z (2011) Nature 480: 372–375
11. Luo X, Park KK (2012) Int Rev Neurobiol 105:141–173
12. Park KK, Liu K, Hu Y, Smith PD, Wang C, Cai B, Xu B, Connolly L, Kramvis I, Sahin M, He Z (2008) Science 322:963–966
13. Park KK, Liu K, Hu Y, Kanter JL, He Z (2010) Exp Neurol 223:45–50

Chapter 18

Time-Lapse In Vivo Imaging of Dorsal Root Nerve Regeneration in Mice

Andrew Skuba, Meredith Ann Manire, Hyukmin Kim, Seung Baek Han, and Young-Jin Son

Abstract

Primary sensory axon injury is common after spinal cord and root injuries and causes patients to suffer chronic pain and persistent loss of sensation and motor coordination. The devastating consequences of such injuries are due primarily to the failure of severed axons to regenerate within the damaged CNS. Our understanding of the molecular and cellular events that play key roles in preventing or promoting functional regeneration is far from complete, in part because complex and dynamic changes associated with nerve injury have had to be deduced from comparisons of static images obtained from multiple animals after their death. Revolutionary innovations in optics and mouse transgenics now permit real-time monitoring of regenerating dorsal root axons directly in living animals. Here, we describe detailed procedures for repetitive monitoring of identified axons in a lumbar dorsal root over hours to weeks using both wide-field and two-photon microscopes. We also discuss the strengths and limitations of in vivo imaging and provide suggestions based on our own experience for troubleshooting issues associated with repeated anesthetization, an extensive laminectomy, and post-op care. These techniques provide the unprecedented opportunity to obtain novel insights into why sensory axons fail to reenter the spinal cord.

Key words Spinal root injury, Rhizotomy, Dorsal root entry zone, Sensory nerve regeneration, In vivo imaging, Dorsal roots

1 Introduction

Following spinal root injury, sensory axons regenerate peripherally in the dorsal root but stop or turn around at their entrance to the spinal cord because of mechanisms that are incompletely understood [1, 2]. The dorsal root entry zone (DREZ), where these axons stop or turn around, represents the junction between the central and peripheral nervous systems. Inhibition by CNS glia such as reactive astrocytes and degenerating oligodendrocytes prevents regeneration at the DREZ, but the decisive factor(s) and their mechanisms of action are unknown.

Andrew J. Murray (ed.), *Axon Growth and Regeneration: Methods and Protocols*, Methods in Molecular Biology, vol. 1162,
DOI 10.1007/978-1-4939-0777-9_18, © Springer Science+Business Media New York 2014

Previous studies relied heavily on conventional tracing techniques and postmortem analyses, which are limited in spatial and temporal resolution, and often provided conflicting or inconclusive evidence. Our ability to monitor neurons serially in vivo has increased dramatically owing to revolutionary innovations in optics and mouse transgenics. These techniques introduce an unprecedented level of dynamic analysis to biological processes such as axon regeneration and dramatically redefine what we can learn directly from living animals. Surprisingly, however, imaging studies of living spinal cord or roots have only recently begun. Despite extensive experience imaging living muscles [3–5] and brain [6–8], the time-lapse imaging of real-time interactions between axons and their environment in the spinal cord or the roots remains a challenge. First, the proximity of the spinal cord to the heart and lungs results in particularly troublesome respiratory-induced movement artifacts. Second, chronic imaging of spinal cord requires multiple invasive surgeries, including laminectomy, with multiple anesthetizations and intubations, and frequent interruption of respiration to acquire stable images.

Over the last few years, pioneering studies from my laboratory have established the feasibility and value of tracking regenerating dorsal root axons in living mice. We have repeatedly monitored the regeneration of identified axons over long distances for hours to months [9–11]. In combination with a targeted ultrastructural analysis, these studies demonstrated that most axons were not repelled but were immobilized rapidly and chronically at the CNS portion of the DREZ, where they formed presynaptic terminal endings. We observed additional important features of axons arriving at the DREZ, which previous static analyses could not discern and which provided unexpected insights that led us to a novel concept ([9–11] and unpublished data).

Repeated in vivo imaging also enables us to avoid mistaking axons that survived lesions for axons that had regenerated. We consider the power to distinguish between spared and regenerating axons to be an important asset of in vivo imaging, especially because axon sparing has been responsible for conflicting observations in studies of spinal cord or root injuries [12].

This chapter provides detailed instructions for performing in vivo imaging of regenerating dorsal root (DR) axons repeatedly in living mice. Our protocol uses H or M line *Thy1*-yellow fluorescence protein (YFP) mice in which a subset of DRG neurons are fluorescently labeled [13]. The genetic tracing of neurons allows repeated imaging of identified DR axons at high spatial and temporal resolution under both widefield and two-photon microscopy. We found that lumbar DRs have a number of advantages over cervical DRs in repeated imaging of identified, regenerating DR axons. The use of inhalation anesthesia has been a recent refinement to our protocol for continuous imaging of DR regeneration over extended periods.

2 Materials

2.1 Animals

1. Male or female, H or M line *Thy1*-YFP mice, aged 2–3 months, weighing 20–30 g (Jackson Laboratory) are used for in vivo imaging experiments (*see* **Note 1**). Animals are maintained under standard care conditions with food and water ad libitum.

2.2 Anesthesia and Postoperative Care

1. Injectable cocktail of xylazine (8 mg/kg) (AnaSed injection sterile solution, Lloyd, Inc.) and ketamine (120 mg/kg) (Ketamine hydrochloride injection USP, Hospira Inc.) in 0.9 % saline or, alternatively, inhalation anesthesia using vaporized isoflurane USP (Piramal Healthcare Limited) in medical oxygen (Airgas) (*see* **Note 2**).
2. Laboratory animal anesthesia system (V-1 LAAS, VetEquip).
3. Injectable Buprenorphine (0.05 mg/kg) (Buprenex injectable, Reckitt Benckiser).

2.3 Surgical Preparation

1. Adjustable temperature controller with heating pad (ATC 1000, World Precision Instruments, Inc.), small animal hair clippers (Oster Professional Products), hair removal lotion (Nair with baby oil, Church & Dwight Co.).

2.4 Surgical Tools and Equipment

1. *Materials and tools*: Gauze sponges (Fisher Scientific), cotton-tipped swabs (Fisher Scientific), 1 mL syringes (BD Biosciences), subcutaneous (Sub-Q) needles (26ga.) (BD Biosciences), spring scissors and forceps (Fine Science Tools), 2.5-mm curved rongeurs (Fine Science Tools), lactated Ringer's injection USP (B. Braun Medical), sterile saline solution (APP Pharmaceuticals), mouse stereotaxic holder with vertebral clamps (SR-AM and STS-A, Narishige).
2. *Operative materials*: Thin synthetic matrix membrane (Bertek Pharmaceuticals), artificial dura (Gore Preclude MVP Dura Substitute, W.L. Gore and Associates), 5-0 Silk sutures (Ethicon Inc.), wound clips (Perfect-Ets Bruneau, Bruneau, France), Kwik-Sil silicone elastomer (World Precision Instruments, Inc.).
3. *Microscopy setups*: For widefield fluorescence microscopy—fluorescent stereomicroscope (M205C, Leica Microsystems), CCD camera (Orca-Rx2, Hamamatsu Corp.), for multiphoton microscopy—(TCS SP5 II, Leica Microsystems), external non-descanned detectors with changeable filter cube for YFP (Leica Microsystems), Tai-sapphire tunable InfraRed laser (Mai-Tai HP, SpectraPhysics).
4. *Acquisition software*: For widefield fluorescence imaging—Metamorph software (Molecular Devices), for multiphoton imaging—Leica Advanced Fluorescence Suite (Leica Microsystems) Photoshop (Adobe).

3 Methods

3.1 Widefield Microscope Setup and Surgical Preparation

1. *Widefield microscope*. Our widefield surgical and imaging setup consists of a fluorescence stereomicroscope (M205C, Leica Microsystems) with a fast shutter and a CCD camera (Orca-Rx2, Hamamatsu Corp.) controlled by Metamorph software (Molecular Devices) (*see* Fig. 1).
2. In advance, adjust a thermostatically controlled heating pad (World Precision Instruments, Inc.) to 32.5 °C. The heating pad is used to maintain a constant body temperature during both the surgical procedure and post-operative care.
3. In advance, warm sterile lactated Ringer's solution (B. Braun Medical) or artificial cerebrospinal fluid (ACSF) in a syringe to 32.5 °C using the heating pad. Warmed physiological solution is used to irrigate the spinal cord during surgery.

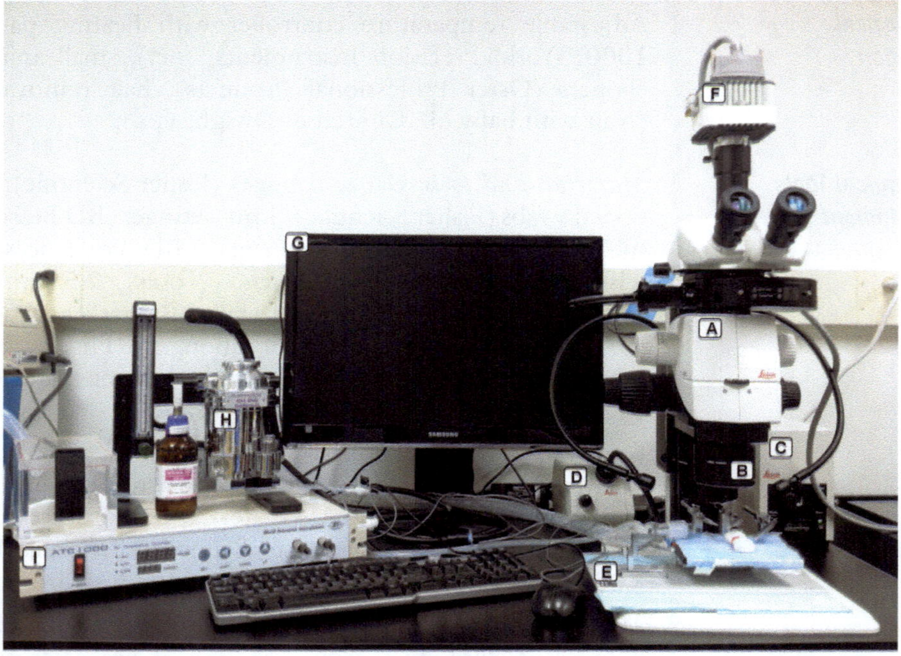

Fig. 1 Setup for surgical procedure and in vivo imaging with a widefield fluorescence microscope. The setup includes (**a**) an upright widefield epifluorescence stereomicroscope (M205C, Leica Microsystems), equipped with (**b**) a revolving turret with ×1 and ×2 objectives (Leica Microsystems), (**c**) a fluorescence light source (Leica Microsystems), and (**d**) brightfield light source (Leica Microsystems). After surgery, the animal is mounted to (**e**) a mouse stereotaxic holder (SR-AM and STS-A, Narishige). Images are acquired using (**f**) a CCD camera (Orca-R2, Hamamatsu Corp.) controlled by (**g**) digital imaging software (Metamorph, Molecular Devices). Throughout the experiment, the animal receives continuous anesthesia from (**h**) a laboratory animal anesthesia system (V-1 LAAS, VetEquip), while body temperature is maintained by (**i**) a heating pad attached to a thermostatically controlled adjustable temperature controller (ATC 1000, World Precision Instruments, Inc.)

4. After weighing the animal, initially anesthetize with an intraperitoneal (i.p.) injection of a cocktail of xylazine (8 mg/kg) and ketamine (120 mg/kg) in 0.9 % saline. Alternatively, inhalation anesthesia using 4 % vaporized isoflurane in 1.0 L/min medical oxygen can be used to initially anesthetize the animal (Laboratory Animal Anesthesia System, V-1 LAAS, VetEquip). Determine the stage of anesthesia and analgesia by assessing the animal's sensitivity by pinching the paw of the hindlimb. The appropriate stage of anesthesia and analgesia must be monitored for the duration of the session.
5. If inhalation anesthesia is used, adjust the dosage (percent isoflurane vaporization in medical oxygen L/min) to 1.50–2.50 % isoflurane vaporization in 1.0 L/min of oxygen for prolonged anesthesia and analgesia (sessions longer than 1 h) (*see* **Note 2**).
6. Once the animal has reached the appropriate stage of anesthesia and analgesia, shave the animal's upper back using small animal clippers. Then, use a cotton-tipped swab to spread one small drop of hair removal lotion over the shaved area. After just a moment, wipe away the lotion using 70 % ethanol-soaked gauze swabs.

3.2 Surgical Exposure of the DREZ and Dorsal Root

1. Gently place the animal on the heating pad (set to 32.5 °C).
2. Using 70 % ethanol-soaked gauze swabs, thoroughly disinfect the skin of the back, ensuring that no stray hairs remain.
3. Position the animal on a support cushion (rolled cotton gauze) to flatten the vertebral column.
4. Under brightfield illumination, use a ten-blade scalpel to make a midline incision (2–3 cm) in the skin of the back. Use cotton-tipped swabs to halt bleeding and to clean the incision.
5. Gently part the spinal musculature above the vertebrae at the midline. Continue until the underlying lumbar vertebrae are exposed. Retraction hooks can be used to hold the musculature and skin open.
6. Using small rongeurs, expose the lumbar level L3–sacral level S1 spinal segments by performing a right-sided hemilaminectomy (*see* **Note 3**). Carefully remove the right dorsal portion of the L5 vertebra at the level of the iliac crest of the hip (location of the L5 DRG) toward the L2 vertebrae (two vertebrae caudal to the last rib). Frequently perfuse the exposed cavity with the warm sterile Ringer's solution or ACSF to prevent the tissue from drying out. The tissue should never be allowed to dry out, as this will damage the spinal cord.

3.3 Dorsal Root Crush Injury/Rhizotomy

1. Before proceeding, ensure that the animal is under deep anesthesia. Assess the animal's sensitivity by pinching the paw of the hind limb before performing the dorsal root crush injury/rhizotomy.
2. At this point in the procedure (approximately 30 min after the first i.p. injection of anesthetic), a supplement (half-dose, 0.5×) should be administered subcutaneously in order to keep the animal fully anesthetized. Alternatively, if inhalation anesthesia is used, 1.50–2.50 % isoflurane in 1.0 L/min of oxygen can provide continuous anesthetization.
3. Cease brightfield illumination, and switch to low-intensity fluorescence excitation, and use an appropriate filter cube to visualize the YFP-labeled (YFP+) axons with the eyepieces (*see* **Note 4**).
4. Use the tip of a Sub-Q (26ga.) needle to make a small incision in the dura overlying the L5 dorsal root. Do not remove the dura at any time during the procedure, and take great care not to damage the underlying spinal cord. Again, perfuse the exposed area repeatedly with warmed sterile physiological solution and clean the area gently with cotton-tipped swabs.
5. Insert one side of a fine forceps (Dumont #5) subdurally, far enough to encompass the medial portion of the L5 root.
6. Close the forceps gently, but firmly, holding the medial portion of the L5 dorsal root for 10 s, and then gently release the forceps.
7. Wash the exposed area repeatedly with warmed sterile physiological solution and clean gently with cotton-tipped swabs.
8. If necessary, gently reposition the dorsal root with forceps so that the site of the crush injury is not obscured.

3.4 Widefield Microscopy Image Acquisition and Postoperative Procedures

1. Using Metamorph or other appropriate controlling software, acquire multiple images of the entire exposed area, including the crush site and the DREZ before and just after the injury. A set of both low- and high-magnification images should be acquired (*see* **Note 5**).
2. Acquire images either as single snapshots or as multiple streams of 10–20 frames acquired within 30–40-ms exposure time. Because the area of interest is quite wide, several in-focus images may be needed for accurate montage reconstruction using Adobe Photoshop (*see* Fig. 3).
3. After imaging is complete, in order to minimize subsequent scar accumulation, tightly apply a piece of thin synthetic matrix membrane (Biobrane) mesh side up. Trim the membrane to an appropriate size so that it will cover the exposed spinal cord. Then place a piece of artificial dura (Gore Preclude) on top.

It may be necessary to slightly bend this thicker membrane so that it follows the natural contour of the vertebrae.

4. Close the musculature with sterile 5-0 sutures (Ethicon). Non-soluble (silk) sutures should be used. Then close the midline skin incision with wound clips (Ets Bruneau, Bruneau, France). Wound clips should not be overly tight.
5. As part of postoperative care, inject Ringer's solution (0.3 to 0.5 mL) subcutaneously near the surgical site on the back of the animal. Administer buprenorphine for postoperative analgesia (0.05 mg/kg) intramuscularly every 12 h for 2 days.
6. Keep the animal on the warmed heating pad (34–35 °C) for several hours until recovered.
7. Make water and food easily accessible to the recovering animal. Hydrogel and chow should be left on the floor of the cage for the animal to easily reach.

3.5 Re-exposure and Repeated Imaging

1. In advance, adjust the temperature of the heating pad's output to 32.5 °C and warm sterile Ringer's solution or ACSF to 32.5 °C.
2. Anesthetize the animal as before, and carefully remove the wound clips and sutures.
3. Gently remove the thick artificial dura, and keep it in a sterile tube containing sterile saline for later reuse. Gently remove the thin synthetic matrix membrane using forceps, and place it in sterile saline.
4. Gently remove accumulated scar tissue with the bent tip of a Sub-Q needle and fine forceps (*see* **Note 6**). Great care should be taken to thoroughly clean the exposed area. Perfuse frequently with warm saline, and halt any bleeding with cotton-tipped swabs.
5. Re-expose the operative field, including L5 DREZ, L5 DR, and the injury site. Under fluorescence excitation, use persistent landmarks to relocate the YFP+ axons imaged in the previous sessions (*see* Fig. 3) (*see* **Note 7**). The branching points of major blood vessels along the main midline vessel and/or fluorescent fragments of degenerating axons can serve as landmarks.
6. If available, mount the animal in an appropriate mouse stereotaxic holder with vertebral clamps (i.e., SR-AM and STS-A, Narishige). The vertebral clamps should be positioned just rostral and caudal to the exposed area and must directly clamp to the vertebrae (*see* Fig. 2). It may be necessary to further reflect the musculature at these regions so that the inset of the clamps makes direct contact with the vertebrae. Secure the animal's head and tail with the proper clamps. Finely adjust the animal's position on the holder (slope and/or rotation) to

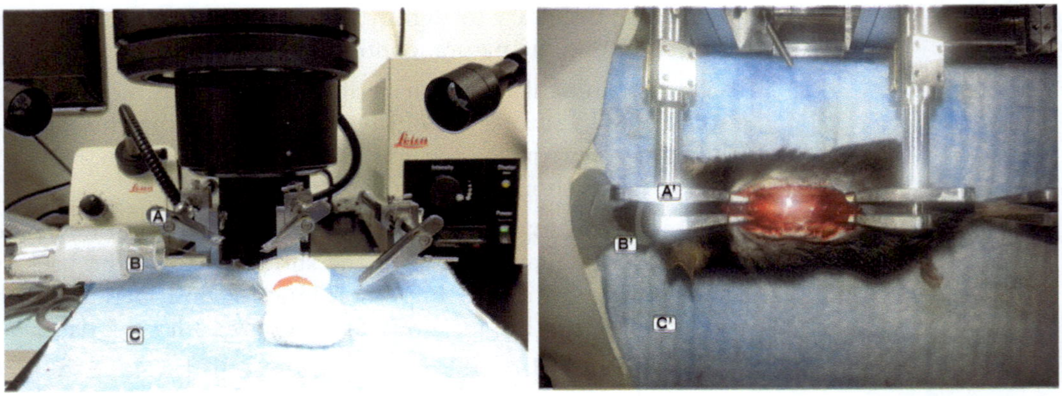

Fig. 2 Animal placement for widefield epifluorescence in vivo imaging. (*Left*) After re-exposing the operative field and cleaning away accumulated scar tissue, support provided by gauze sponge "pillows" can be removed and the animal mounted to (A, A′) the manipulators of the stereotaxic holder using the headmount (*leftmost*), the two vertebral clamps (*center*), and the tail suspender (*rightmost*) (SR-AM and STS-A, Narishige) to reduce movement of the animal due to respiration. The vertebral clamps are positioned just rostral and caudal to the exposed area and clamp directly to the vertebrae. During the imaging session, (B, B′) a nosecone attached to a laboratory animal anesthesia system (V-1 LAAS, VetEquip) delivers continuous inhalation anesthesia (Isoflurane, Piramal Health Limited) in medical oxygen (Airgas), while the animal's body temperature is maintained by (C, C′) a heating pad attached to a temperature controller (ATC 1000, World Precision Instruments, Inc.). The exposed area is frequently perfused with warmed sterile Ringer's solution

accurately reproduce the animal's orientation during the previous imaging sessions (*see* **Note 8**).

7. The animal should be closely monitored and checked with paw pinch to ensure proper maintenance of anesthesia and analgesia. A supplementary (0.5× dose) anesthetic should be administered as needed.
8. As in previous imaging sessions, several in-focus low-magnification images of the L5 DREZ and L5 DR, including the crush site, should be acquired, with persistent landmarks clearly visible in each acquired image. Higher magnification images should also be acquired, despite the narrow field of view, for improved details (*see* Fig. 3).

3.6 Multiphoton Microscope Setup

1. Our multiphoton imaging setup consists of a Leica TCS SP5II microscope optimized for both confocal and multiphoton imaging controlled by Leica LAS AF software (Leica Microsystems) (*see* Fig. 4).
2. At least 1 h prior to imaging, ready the microscope and tunable IR laser (Mai-Tai HP, SpectraPhysics) for imaging. Start up the acquisition software for multiphoton image acquisition, and activate the resonance scanner if equipped.
3. Keep the animal on the heating pad set to 32.5 °C, and, if available, properly position the animal (still mounted to the stereotaxic holder) on the lowered microscope stage.

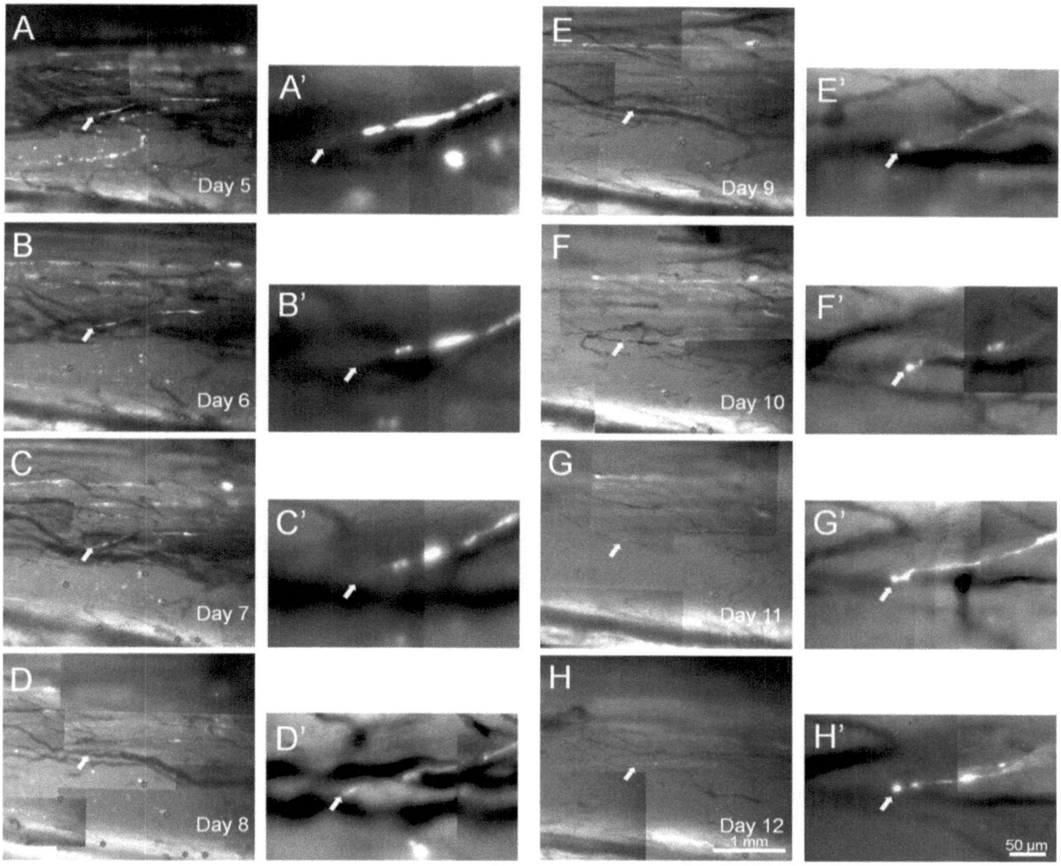

Fig. 3 Re-identification and repeated widefield in vivo imaging of regenerating L5 dorsal root axons. A YFP+ axon at the DREZ was imaged on days 5, 6, 7, 8, 9, 10, 11, and 12 days after L5 root crush. Over the course of several days after root crush, axons distal to the lesion are fragmented and very slowly degraded in the CNS territory of the DREZ. These fluorescent axon fragments, as well as branches of blood vessels, can be used as landmarks to relocate axons imaged in previous sessions (*asterisks*). The axon tip (*arrow*) is magnified in the lower panels (composite images of in-focus frames using Adobe Photoshop)

4. If using a dipping objective (e.g., 20× immersion, 507701, Leica Microsystems), create a well encompassing the exposed area using Kwik-Sil (World Precision Instruments, Inc.) as a barrier; apply several layers of elastomer, and allow them to dry for at least 10 min. Fill this well with warmed sterile Ringer's solution, and lower the objective into focus (*see* Fig. 4).
5. Using low-intensity fluorescence excitation, an appropriate filter, and the eyepieces, adjust the focus knob (z-position) so that the YFP+ axons are clearly in focus.
6. Move the stage in either an *X* or a *Y* direction so that the axon ending is properly in the field of view. At this point, slight adjustments to the animal's orientation may be necessary to reproduce the field of view from the previous imaging session.

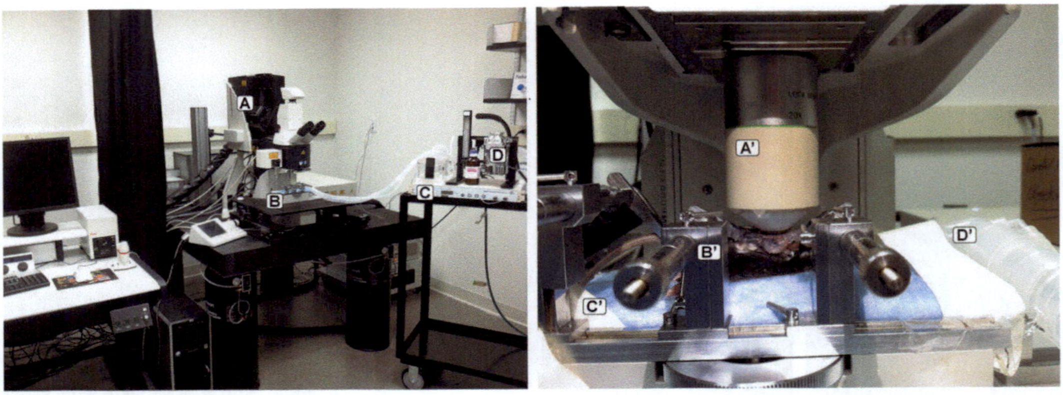

Fig. 4 Setup for in vivo imaging with a multiphoton microscope. The setup includes (**a**) an upright microscope optimized for confocal and multiphoton in vivo imaging (TCS SP5 II, Leica Microsystems), controlled by imaging software (LAS AF, Leica Microsystems) with (**a**′) a ×20 dipping objective (×20 water immersion, 507701, Leica Microsystems). The animal mounted to (**b**, **b**′) the mouse stereotaxic holder (SR-AM and STS-A, Narishige) is carefully positioned to reproduce the orientation in previous imaging sessions. A Kwik-Sil (World Precision Instruments, Inc.) "well" encompassing the exposed area is filled with warmed sterile Ringer's solution. (**c**) A temperature controller (ATC 1000, World Precision Instruments, Inc.) and (**c**′) attached heating pad maintain the animal's body temperature through the experiment. For prolonged (>1 h) sessions, continuous inhalation anesthesia is consistently delivered via (**d**) a laboratory animal anesthesia system (V-1 LAAS, VetEquip) through (**d**′) an attached nosecone. A custom curtain is used to block external light from reaching the microscope

3.7 Multiphoton Image Acquisition

1. In *XYZ* scanning mode, set the format of the image to at least 512 × 512 format.
2. If not equipped with a resonance scanner, choose a scanning speed (in Hz) that balances faster scanning speed without adding unnecessary noise.
3. Tune the IR laser to the appropriate wavelength (~890 nm for YFP) for excitation. Laser power (or intensity) should be kept low to avoid photobleaching or phototoxicity.
4. Activate non-descanned detectors (NDDs) or appropriate detectors for YFP.
5. While minimizing the duration of scans, adjust the gain and offset appropriately. Adjust the laser intensity if necessary.
6. For image acquisition, set up a z-stack (z-series) that contains the axon ending of interest (~20–50 μm). Set the step size to 1 μm or less (depth between image planes). Finally, adjust the format of the image to at least 1,024 × 1,024 and acquire a z-stack with line average (from 1 to 4) to reduce noise (*see* Fig. 5).
7. After z-stacks have been acquired, alignment software (such as AutoAligner, BitPlane) can be used to manually align off-centered optical sections.
8. Image acquisition can be repeated with an increased digital zoom (1–3) or at higher magnification if available.

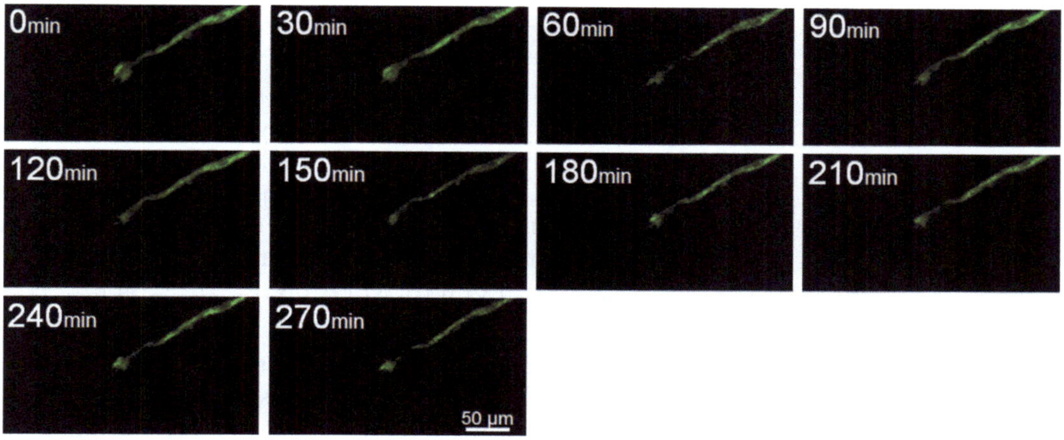

Fig. 5 An example of time-lapse multiphoton in vivo imaging of a DR axon that has reached the DREZ after L5 dorsal root crush. $T = 0$ min, an axon that has reached the DREZ. The tip of this axon remains in the same location for more than 4.5 h of time-lapse imaging (maximum projections of z-series)

4 Notes

1. Regarding a mouse model for in vivo imaging, we utilize transgenic mice expressing YFP under the control of the *Thy1* promoter [13]. These mice take advantage of *Thy1* as a neuron-specific promoter to express YFP in the nervous system. Among the diverse lines generated by Feng et al., the M line of *Thy1*-YFP mice (YFP-M) is ideal for in vivo imaging of sensory axons as only a few (<10 %) dorsal root ganglion (DRG) neurons express YFP. This unique pattern allows several distinct, superficially positioned YFP+ axons to be monitored simultaneously. We have previously determined that these axons belong to the large-diameter, neurofilament-positive (NF+) proprioceptive subpopulation of DRG neurons. In addition, although variable among littermates, the separation distance between adjacent labeled axons allows individual axons to be re-identified and monitored individually. The trajectory of intact axons as they enter the spinal cord is stereotypical. Also, because few axons express YFP, Wallerian degeneration produces only a small number of fragments, which facilitates identification of an axon ending.
2. After using the aforementioned injectable anesthetic cocktail for years, we have recently found that properly ventilated inhalation anesthesia using vaporized isoflurane in medical oxygen more consistently produces a constant level of anesthesia and analgesia, which we monitor throughout the experiment. In addition to providing more rhythmic breathing than injectable anesthesia, inhalation anesthesia avoids the need to change

the animal's position during imaging in order to administer supplemental doses.

3. We have learned from experience that superficially positioned YFP+DR axons approaching and arriving at the DREZ are best imaged in the lumbar spinal cord. A laminectomy extending from level L3–sacral level S1 spinal segments will expose DR axons running along the length of spinal cord, permitting easy imaging of their peripheral regeneration.
4. Intact/non-injured YFP-labeled (YFP+) axons in M line *Thy1*-YFP mice appear wide and brightly labeled. Superficially positioned intact YFP+axons do not require "full-intensity" fluorescence excitation to reliably visualize or even image them. Importantly, the intensity should be kept at minimum to avoid photobleaching and phototoxicity. These cautions apply even more strongly to imaging the thin, dimly labeled YFP+axons that approach and arrive at the DREZ after injury.
5. During widefield epifluorescence microscopy and imaging of the exposed spinal cord, including the DREZ, DR, and injury site, we found it advantageous to acquire sets of images at both low and high magnification. Several low-magnification images of the most sharply focused fields of view can later be combined into montages that display the entire exposed area. However, because they lack resolution, low-magnification images must be supplemented with corresponding higher magnification images that offer additional details of a circumscribed field of view.
6. Repeated surgical re-exposure of the area of interest will cause gradual formation of a fibrotic scar that encompasses the exposed spinal cord and requires careful cleaning and removal. To minimize scar formation around the spinal cord, we tightly cover the exposed spinal cord and dura with pieces of thin synthetic matrix membrane and artificial dura after every imaging session. The goal is for the scar tissue to accumulate on the membrane rather than on the spinal cord, which will allow it to be removed when the membranes are taken out. However, despite the use of these membranes, scar accumulates and must be cleaned by a tedious process that risks damage to the dura. Incomplete removal may cause image distortion that makes analysis more difficult.
7. To re-identify individual axons in subsequent sessions, several landmarks should be used, including bright fragments of degenerating axons that remain up to several days after injury, spared or intact axons, and branching points of several blood vessels off the prominent dorsal vein. It is important to be able to re-identify a regenerating axon that has grown more deeply

into the surrounding tissues between sessions. In such cases, if it is available, multiphoton microscopy can be used to visualize deeper tissues with greater spatial resolution.

8. The animal's orientation to the objective lens should be consistently kept near to perpendicular. Maintaining this orientation may prove challenging because the axons do not course through the root or into the spinal cord in a single plane parallel to the dorsal surface. To meet this challenge, we have found that cotton gauze can be rolled and taped to form "pillows" which, when used in conjunction with adjustable retraction hooks, can enable the experimenter to position the animal properly. We have found that subsequent imaging sessions demand reproducibility of the animal's orientation under the objective lens. Even the natural rhythmic movement of respiration represents an additional obstacle to maintaining proper orientation. We have found it helpful to mount the animal directly to a stereotaxic holder (Narishige) that is designed for this purpose. This device improves orientation reproducibility in repeated imaging sessions.

Acknowledgement

We thank Drs. Tim Himes and Alessandro Di Maio for earlier contribution to imaging dorsal root axons in living mice. Our work was supported by grants from NIH (NS079631, NS062320), Shriners Hospitals for the Children, and Craig Neilsen Foundation.

References

1. Smith GM, Falone AE, Frank E (2012) Sensory axon regeneration: rebuilding functional connections in the spinal cord. Trends Neurosci 35(3):156–163
2. Han SB, Kim H, Skuba A, Tessler A, Ferguson T, Son YJ (2012) Sensory axon regeneration: a review from an in vivo imaging perspective. Exp Neurobiol 21(3):83–93
3. Gan WB, Bishop DL, Turney SG, Lichtman JW (1999) Vital imaging and ultrastructural analysis of individual axon terminals labeled by iontophoretic application of lipophilic dye. J Neurosci Methods 93(1):13–20
4. Bishop DL, Misgeld T, Walsh MK, Gan WB, Lichtman JW (2004) Axon branch removal at developing synapses by axosome shedding. Neuron 44(4):651–661
5. Lu J, Lichtman JW (2007) Imaging the neuromuscular junction over the past centuries. Sheng Li Xue Bao 59(6):683–696
6. Zuo Y, Yang G, Kwon E, Gan WB (2005) Long-term sensory deprivation prevents dendritic spine loss in primary somatosensory cortex. Nature 436(7048):261–265
7. Trachtenberg JT, Chen BE, Knott GW, Feng G, Sanes JR, Welker E, Svoboda K (2002) Long-term in vivo imaging of experience-dependent synaptic plasticity in adult cortex. Nature 420(6917):788–794
8. Pan F, Gan WB (2008) Two-photon imaging of dendritic spine development in the mouse cortex. Dev Neurobiol 68(6):771–778
9. Di Maio A, Skuba A, Himes BT, Bhagat SL, Hyun JK, Tessler A, Bishop D, Son YJ (2011) In vivo imaging of dorsal root regeneration: rapid immobilization and presynaptic differentiation at the CNS/PNS border. J Neurosci 31(12):4569–4582
10. Hanna AS, Son YJ, Dempsey R (2011) Live imaging of dorsal root regeneration and the

resurgence of a forgotten idea. Neurosurgery 69(2):N18–N21

11. Skuba A, Himes BT, Son YJ (2011) Live imaging of dorsal root axons after rhizotomy. J Vis Exp 55:e3126

12. Steward O, Zheng B, Tessier-Lavigne M (2003) False resurrections: distinguishing regenerated from spared axons in the injured central nervous system. J Comp Neurol 459(1):1–8

13. Feng G, Mellor RH, Bernstein M, Keller-Peck C, Nguyen QT, Wallace M, Nerbonne JM, Lichtman JW, Sanes JR (2000) Imaging neuronal subsets in transgenic mice expressing multiple spectral variants of GFP. Neuron 28(1):41–51

Chapter 19

Using Manganese-Enhanced MRI to Assess Optic Nerve Regeneration

Ioanna Sandvig and Axel Sandvig

Abstract

Experimental visual pathway lesion in the form of optic nerve (ON) crush or transection injury results in massive death of retinal ganglion cells (RGCs) and permanent loss of synaptic connections (Berkelaar et al., J Neurosci 14:4368–4374, 1994). Despite the fact that RGC axon regeneration is inhibited in a manner typical of other CNS lesions, the rodent ON injury model is one of the few models where robust axon regeneration has been achieved after therapeutic intervention (Berry et al., Restor Neurol Neurosci 26:147–174, 2008). However, assessment of the efficacy of therapeutic approaches in promoting ON regeneration has traditionally relied on histological methods, which necessitate the sacrifice of experimental animals and thus preclude longitudinal in vivo monitoring of individual subjects. Manganese-enhanced MRI (MEMRI) utilizes the paramagnetic properties and uptake and transport mechanisms of manganese ions (Mn^{2+}) by neurons, thus enabling serial in vivo monitoring of the entire axonal projections (Sandvig et al., J Magn Reson Imaging 34:670–675, 2011; Thuen et al., J Magn Reson Imaging 4:492–500, 2005; Pautler et al., Magn Res Med 50:33–39, 2003; Saleem et al., Neurotechnique 34:685–700, 2000). The above properties of Mn^{2+} render MEMRI a highly suitable technique for assessment of ON regeneration after injury, especially with a view to in vivo monitoring of neuronal connectivity and axon-regenerative responses to treatment. In this chapter, we provide a generic protocol for ON lesioning and MEMRI application for assessment of ON regeneration in rodents.

Key words MEMRI, Contrast agent, Visual pathway, Axotomy, Central nervous system, Connectivity

1 Introduction

Experimental visual pathway lesion is an excellent model for the study of CNS damage and repair both in terms of clinical relevance and neuroanatomical properties. Direct trauma or ischemia as well as different types of chronic ophthalmic conditions related to diseases such as glaucoma and diabetes induce severe visual impairment as a result of extensive RGC loss [1, 2]. Furthermore, demyelinating ON pathologies often constitute early symptoms of various inflammatory and neurodegenerative diseases such as multiple sclerosis [1]. With regard to experimental CNS damage and repair, special neuroanatomical features of the visual system,

Andrew J. Murray (ed.), *Axon Growth and Regeneration: Methods and Protocols*, Methods in Molecular Biology, vol. 1162,
DOI 10.1007/978-1-4939-0777-9_19, © Springer Science+Business Media New York 2014

i.e., the exclusively centripetal direction of the RGC axon projection and relative accessibility of the vitreous body and retina, present distinct advantages in such investigations as they facilitate controlled lesioning as well as therapeutic and imaging strategies which interrogate one axonal pathway [3–6].

The central visual system is part of the brain and consists of the eye, retina, and retinofugal projection [3, 7]. While RGC somata are located exclusively in the retina, RGC axons exit the retina through the ON head, become myelinated at the lamina cribrosa, and form bundles inside the ON, which passes through the orbit and optic foramen into the cranial cavity [3, 7]. In rodents and lower vertebrates, 90 % of the axons of the left and right ON decussate at the chiasm, immediately rostral to the pituitary gland, and form the optic tracts, which subsequently project to the lateral geniculate nucleus (LGN), superior colliculus, and midline pretectal area and then into the visual cortex in the occipital lobe [3].

Visual pathway lesion can be induced at different distances from the lamina cribrosa by means of optic nerve crush (ONC) or transection (ONT) [8–11]. Following intraorbital axotomy, massive RGC loss (85–90 %) occurs within the first 2–3 weeks post-lesioning, effectively severing neuronal connections with the contralateral LGN and superior colliculus [7, 12, 13]. In adult mammals, spared RGC axons initially respond to the injury by entering a program of abortive regeneration during which axonal sprouting is observed in the proximal ON segment, i.e., from the lamina cribrosa to the injury site [3]. However, in the absence of therapeutic treatment, sprouting RGC axons fail to traverse the non-permissive lesion microenvironment and, as a result, axonal growth is arrested by 20 days post-lesioning (dpl) [3].

Despite the fact that ON regeneration is inhibited by a host of intrinsic and extrinsic factors characteristic of CNS lesion pathology [14–17], a number of therapeutic interventions may promote axon regeneration beyond the lesion site and into the distal stump [18–20]. Such approaches include intravitreal (*ivit*) [4, 20–22] and intra-optic nerve (*iON*) engraftment of peripheral nerve segments [23–25], *ivit* administration of neurotrophic factors [19], activation of inflammatory cytokines [4, 8, 18, 26–30], activation of the mTOR pathway [31, 32], as well as *iON* engraftment of olfactory ensheathing cells [6, 33–36].

The success of therapeutic strategies in promoting ON regeneration largely depends on the extent to which they induce sustained axonal growth both proximal and, importantly, distal to the lesion site. Thus, the ability to visualize the optic nerve in vivo significantly enhances the rigor of such investigations by enabling longitudinal monitoring of axon-regenerative responses to treatment and correlative study of functional recovery with anatomical restitution [5, 6, 37, 38]. The MEMRI technique, which utilizes Mn^{2+} as an in vivo tract tracer, is highly suitable for this purpose as it circumvents the limitations of traditional tracers, such as carbocyanine fast Di-I and FluoroGold, by allowing the use of live subjects for serial noninvasive neuroanatomical imaging.

MEMRI of the visual pathway relies on two main attributes of Mn^{2+}: (1) their paramagnetic properties, which shorten the spin–lattice relaxation time of surrounding water protons and, as a result, produce positive contrast with T1-weighted MRI in tissue areas where Mn^{2+} are accumulated [39, 40], and (2) the fact that being a Ca^{2+} analogue, Mn^{2+} are taken up by neurons through L-type voltage-gated Ca^{2+} channels [41, 42]. Upon entry into the neuron, Mn^{2+} are sequestered in the endoplasmic reticulum and subsequently transported along axonal microtubules via kinesin-mediated anterograde transport mechanisms [38, 43, 44] and trans-synaptically relayed to the next neuron in the circuit [4, 5, 45, 46]. Thus in intact animals, direct administration of Mn^{2+} to RGC somata by means of *ivit* injection of aqueous Mn^{2+} solution results in enhancement of the retina and the entire visual projection under T1-weighted MRI [4, 5, 38, 43, 47]. On the other hand, following ON lesion, the Mn^{2+}-induced hyperintense signal stops at the lesion site, effectively delineating the lesion [4–6, 37], thus confirming the hypotheses that live axons are required for Mn^{2+} transport and that Mn^{2+} movement within neurons is via active transport, rather than passive diffusion [4, 6, 38]. This suggests that potential increases in the Mn^{2+}-induced signal proximal and/or distal to the ON lesion site will be consistent with axon-regenerative responses to treatment, rendering MEMRI a superior tract tracing method for longitudinal monitoring and assessment of the efficacy of therapeutic interventions after ON injury [5, 6, 37, 38, 48, 49] as well as studies of comparative physiology between regenerating and non-regenerating species [4].

Despite the many advantages of MEMRI, the risk of Mn^{2+}-induced toxicity as a result of repeated/high doses of $MnCl_2$ constitutes a main caveat in investigations utilizing Mn^{2+} as an in vivo tract tracer [50]. However, for application in the rodent visual pathway, *ivit* doses of 150–300 nmol $MnCl_2$ are safe and yield optimal ON enhancement at 7T [38], while they may also have a neuroprotective effect on RGCs, possibly through competing with Ca^{2+} influx and, effectively, reducing excitotoxicity [38]. It has also been shown that there is a plateau of Mn^{2+} uptake by RGC neurons after *ivit* $MnCl_2$ injection [43]. This suggests that optimal ON contrast enhancement is contingent on prolonged availability rather than high dosage of *ivit* Mn^{2+} [43]. Based on this principle, we have previously demonstrated the utility of biopolymers such as modified Mn^{2+}-alginate microbeads as systems for controlled release of *ivit* Mn^{2+} [51]. Such systems can fulfil the dual purpose of optimizing ON contrast enhancement and, at the same time, minimizing the risk of Mn^{2+}-induced cytotoxicity [51, 52]. Further tailoring of these systems, which address specific biophysical properties of RGC axons at different stages of maturity and regeneration in terms of Mn^{2+} uptake, transport, and clearance, may provide significant methodological refinements for safe application of the MEMRI technique in longitudinal studies of ON regeneration.

The following section details the techniques involved in inducing visual pathway lesion and applying MEMRI to assess ON regeneration in rodents. A protocol for controlled release of *ivit* Mn^{2+} using Mn^{2+}-alginate microbeads is also provided.

2 Materials

2.1 Visual Pathway Lesion

1. Dissecting microscope.
2. Rat head holder with adjustable bite bar and ear bars.
3. Heating pad.
4. Large scalpel blade (No. 22).
5. Pair of spring scissors with fine tip.
6. Pair of sharp scissors with curved blades.
7. Microforceps with fine tip (e.g., Dumont No. 5).
8. Forceps with less fine tip for clamping (e.g., Dumont No. 44).
9. Spongostan.
10. Sterile cotton swabs.
11. Rolled tissue paper bits (approximate length 10 mm; diameter 2 mm, *see* **Note 1**).
12. Rat hair clipper.
13. Disinfecting solution (alcohol, Betadine).
14. 5-0 nylon/silk sutures.
15. Eye lubricant (Simplex ointment, Viscotears).
16. Anaesthetic (fentanyl-fluanisone/midazolam).
17. Analgesic (buprenorphine).

2.2 Mn^{2+} for Tract Tracing of the Visual System

2.2.1 Ivit Injection of Aqueous Mn^{2+} Solution

1. $MnCl_2$ (150–300 nmol in sterile H_2O; injection volume 3 μl).
2. Rat head holder with adjustable bite bar and ear bars.
3. Dissecting microscope.
4. Heating pad.
5. Pulled-glass capillary micropipette with a tip diameter of ~0.1–0.2 mm (or 25G needle, *see* **Note 2**).
6. Capillary tube puller.
7. Capillary glass (borosilicate, outer diameter (OD) 1.5 mm, inner diameter (ID) 0.86 mm, length (L) 100 mm).

8. Fine long pipette tips (10 μl) for backfilling of glass pipettes (*see* **Note 2**).
9. Polyethylene tubing (0.4 mm diameter) attached to a 50 ml sterile syringe (or 5 μl Hamilton syringe, *see* **Note 2**).
10. Pipettor (1–10 μl, *see* **Note 2**).
11. Curved forceps.
12. Anaesthetic (isoflurane or fentanyl-fluanisone/midazolam combination, *see* **Note 3**).

2.2.2 Ivit Injection of Mn^{2+}-Alginate Microbeads

1. $MnCl_2$ (150–300 nmol in sterile H_2O; injection volume 3 μl).
2. Rat head holder with adjustable bite bar and ear bars.
3. Dissecting microscope.
4. Heating pad.
5. Pulled-glass capillary micropipette with a tip diameter of ~0.1–0.2 mm (or 25G needle, *see* **Note 2**).
6. Capillary tube puller.
7. Capillary glass (borosilicate, OD 1.5 mm, ID 0.86 mm, L 100 mm).
8. Fine long pipette tips (10 μl) for backfilling of glass pipettes (*see* **Note 2**).
9. Polyethylene tubing (0.4 mm diameter) attached to a 50 ml sterile syringe (or 5 μl Hamilton syringe, *see* **Note 2**).
10. Pipettor (1–10 μl, *see* **Note 2**).
11. Curved forceps.
12. Anaesthetic (isoflurane or fentanyl-fluanisone/midazolam, *see* **Note 3**).

3 Methods

3.1 Visual Pathway Lesion

Visual pathway lesion can be induced either intraorbitally or intracranially [3]. Intraorbital lesion by means of ONC or ONT 2 mm from the lamina cribrosa is widely used in studies of ON regeneration [3]. Correctly performed ONC, as described by Berry et al. [21], will leave the central retinal artery intact but will result in 85–90 % RGC loss and effectively sever connections with the contralateral visual cortex [3]. As an alternative to ONC, ONT can be performed using the same surgical protocol to access and transect the ON instead of crushing it [3, 4]. ONT may be the preferred lesioning method in investigations which aim to demonstrate complete RGC axon regeneration in spontaneously regenerating species such as lower vertebrates [4]. ONT may also be used when the therapeutic intervention involves bridging the lesion by inserting peripheral nerve grafts between the proximal and distal ON stumps [20, 21]. Here we provide a generic protocol for ONC injury in rodents as described by Berry et al. [21].

1. Sedate animal with 3.5–4 % isoflurane.
2. Anaesthetise animal (fentanyl-fluanisone/midazolam, subcutaneously).
3. Confirm lack of toe pinch reflex.
4. Administer analgesic (buprenorphine, subcutaneously).
5. Shave animal head up to the neck.
6. Place animal on heating pad.
7. Position anaesthetized animal in rat head holder, and level the head properly to allow visual inspection and undisturbed access to the eye.
8. Adjust microscope.
9. Swab animal head with alcohol.
10. With scalpel, make 1.5–2.0 cm long midline incision through the skin on top of the head. Avoid cutting through the periost.
11. Make blunt dissection of subcutaneous fascia to expose the temporal ridge, anterior margin of the temporal muscle, posterior orbital margin, superior orbital vein, and supraorbital neurovascular bundle (SONVB).
12. With fine scissors, make a vertical incision in the temporalis fascia 2–3 mm posterior to the periorbital vein (POV), inferior to the temporal ridge.
13. With left hand, using fine forceps, grab hold of the anterior incised margin of the temporal fascia and pull the fascia down and away from the POV.
14. With right hand, using the tips of fine surgical scissors with curved blades facing up towards the superior orbital margin and above the POV, make small stepwise incisions just below the superior orbital margin, ensuring that the latter does not get pierced. For each incision, move the forceps to grab hold of the fascia just posterior to the next cut and pull the fascia down and away from the POV and superior orbital margin.
15. Continue until you reach the supraorbital nerve (SON), beyond which the cut should be angled laterally.
16. With left-hand forceps, make a punch hole into the left side of the SONVB, taking care not to pierce any of the vascular structures, thus avoiding hemorrhage.
17. With right-hand forceps, punch a hole into the right side of the SONVB medial to the left-hand forceps and pinch both forceps hard for 30 s before releasing.

18. With right-hand small surgical scissors, cut the SONVB holding within the supraorbital fascia. The superior lobe of the Harderian gland (SLHG) will come into view.
19. Starting from the defect in the fascia, push the closed forceps blades down laterally to the SLHG, splitting the superior orbital fascia (SOF) posteriorly up to the temporalis muscle. Now the entire SLHG will come into view.
20. With right-hand forceps, grab hold of the SLHG and pull it anteriorly.
21. Identify the lateral and superior rectus muscles and the deep lobe of the Harderian gland (DLHG) between the two muscles.
22. Split the capsule of the DLHG with forceps, grab hold of the DLHG, and pull it anteriorly. The ON can now be visualized beneath the retractor bulbi muscle (RBM) and the dural sheath.
23. Place rolled tissue paper bits to act as retractors between the (1) ON and superior rectus (SR) and (2) ON and lateral rectus (LR), and split the RBM to expose the dura.
24. Grab hold of the dura with fine forceps and incise longitudinally, avoiding damage to the retinal artery and vein running in the medial wall of the dural sheath.
25. On the medial margin of the ON, identify the retinal artery and vein.
26. With left-hand forceps, grab hold of the lateral margin of the split dural sheath and lift gently.
27. With right-hand broad forceps, pinch across the ON, 2 mm distal from the lamina cribrosa, and hold for 10–20s, ensuring that the retinal artery remains intact during this process.
28. Remove the paper bits.
29. Place the inferior Harderian gland underneath the lateral rectus muscle by lifting the latter and pulling it medially over the inferior section of the Harderian gland.
30. Place the superior part of the Harderian gland by pulling it posterior to its normal anatomic space.
31. Suture the skin flap with four single sutures using 4.0 monofilament.
32. Lubricate the eye (Simplex ointment, Viscotears).
33. Return the animal to cage, and monitor recovery from surgery.

3.1.1 Expected Results

The above procedure should not affect the integrity of the eye although slight ischemia may be temporarily observed at the end of the surgery. However, successful lesioning should avoid damage to the central retinal artery and the eye should remain vascularized post-lesioning. Similarly, development of pathologies such as eye

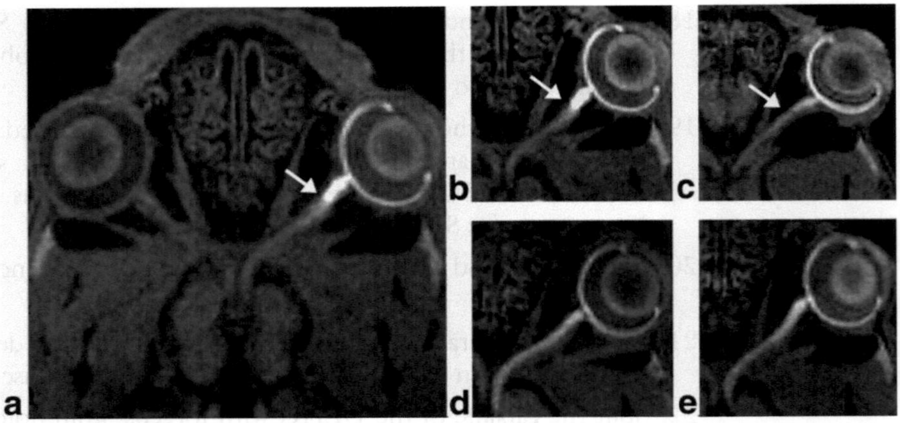

Fig. 1 MEMRI of ONC-lesioned (**a–c**) and unlesioned control rats (**d**, **e**) at 1 and 20 dpl, 24 h after *ivit* $MnCl_2$ injection. The Mn^{2+}-enhanced signal clearly delineates the lesion at 1 dpl (**a**, **b**), and 20 dpl (**c**); however, the signal intensity at the proximal stump (i.e., from the lamina cribrosa to the lesion site) is visibly diminished at the latter time point (**c**). In unlesioned controls, clear Mn^{2+}-induced enhancement of the ON can be observed from the lamina cribrosa to the chiasm at both 1 dpl (**d**) and 20 dpl (**e**). Reprinted from Thuen et al. [5], copyright 2005, with permission from Wiley

cataracts will indicate that the surgery was not performed properly. Successful ONC can be verified by delineation of the lesion site when MEMRI is performed acutely (i.e., 24–48 h) post-lesioning, as illustrated in Fig. 1.

3.2 Tract Tracing of the Visual System

3.2.1 Ivit Injection of Aqueous Mn^{2+} Solution

1. Sedate animal with 3.5–4 % isoflurane (*see* **Note 3**).
2. Anaesthetize the animal (fentanyl-fluanisone/midazolam, subcutaneously, *see* **Note 3**).
3. Confirm lack of toe pinch reflex.
4. Place animal on heating pad.
5. Position the anaesthetized animal in rat head holder, and tilt the head to allow visual inspection and undisturbed access to the eye.
6. Adjust microscope.
7. Check the integrity of the tip of the pulled-glass micropipette under the microscope.
8. Backfill the glass micropipette with 3 μl 150 nmol $MnCl_2$ using a 1–10 μl pipettor and fine long pipette tips.
9. Connect the glass micropipette using polyethylene tubing to a 50 ml sterile syringe in which the plunger has been partially withdrawn (*see* **Note 2**).
10. Expose the eye by slightly pinching the skin or by carefully placing the eye bulb between the open tips of curved forceps and gently pushing the tips down.

11. Insert the tip of the glass micropipette into the vitreous body immediately posterior to the ora serrata, and hold it steady.
12. Using the other hand, slowly inject the $MnCl_2$ by pressing the syringe plunger while monitoring the advancement of the meniscus inside the glass micropipette. Care should be taken not to press air through the micropipette at the end of the injection (*see* **Note 2**).
13. Leave the tip of the glass micropipette inside the vitreous for at least 30 s to minimize reflux.
14. Slowly withdraw the glass micropipette.
15. Return the animal to cage, and allow it to recover from anaesthesia.
16. Check the integrity of the eye prior to imaging.

3.2.2 Animal Preparation for MRI

1. Sedate animals with 4 % isoflurane.
2. Lubricate the eyes (Simplex ointment, Viscotears).
3. Intubate the animal inside the scanner, and maintain ventilation with 2 % isoflurane in O_2.
4. Use ear bars to stabilize the animal, and reduce head movement.
5. Adjust surface coil.
6. Maintain body temperature inside the scanner with circulating water under the animal bed heated at 37 °C.
7. Monitor body temperature, blood pressure, and respiration rate.

3.2.3 Imaging Parameters

Imaging parameters will need to be optimized according to the experimental protocol, type of head surface coil, and magnetic field strength. The following imaging parameters are provided as a guideline for in vivo MEMRI of the ON at 7T, using a T1-weighted 3D FLASH gradient echo sequence [6]:

TR = 12 ms.

TE = 3.5 ms.

Flip angle = 30 %.

FOV = 3.5×2.2 cm^2.

Acquisition matrix = $224 \times 192 \times 141$.

Voxel resolution = $156 \times 156 \times 156$ μm/voxel.

NEX = 12.

Acquisition time = 33 min 38 s.

(*See* **Note 4**).

3.2.4 MEMRI Data Analysis

To determine ON regeneration in response to treatment, the MEMRI data obtained from the T1-weighted 3D FLASH volumes should be analyzed by (1) calculating contrast-to-noise ratios

(CNR) between the Mn^{2+}-enhanced and non-enhanced ON in each animal and (2) comparing the CNR profiles along the ON of treated versus untreated animals and performing appropriate statistical analyses [4–6, 37, 38]. CNR are calculated from the formula $CNR = 0.655(S_{mn} - S_0)/SD_{air}$, where S_0 represents the signal intensities in a region of interest (ROI) in the non-enhanced ON, and SD_{air} is the mean value of the SD in two ROI in air. Elevated CNR profiles in the proximal and/or distal ON segment of treated animals are consistent with RGC axon-regenerative responses to treatment [4–6]. Additionally, histological data can be used for further verification of the MEMRI findings. ROI can be defined at different distances along the ON; however, performing comparisons of isotopic Mn^{2+}-enhanced ON segments at 0–2 mm, 0–4 mm, 1–3 mm, and 2–3 mm from the lamina cribrosa, as well as of intermediate segments within the designated ON segments above, is recommended [6]. In our studies, we have performed the above analyses using the semiautomatic ON segmentation technique described by Olsen et al. [52] (*see* **Note 5**).

Expected Results

Following successful $MnCl_2$ injection, the vitreal humor should appear bright while the retina and ON should be clearly enhanced using T1-weighted imaging. In intact animals, Mn^{2+}-induced signal enhancement should be observed in the entire visual projection, i.e., from the lamina cribrosa, to the LGN, and superior colliculus, and into the visual cortex in the occipital lobe (Fig. 2). In lesioned animals, ONC injury should be clearly delineated as the Mn^{2+}-enhanced signal will stop 2 mm caudal to the lamina cribrosa (Fig. 1). Verification of the lesion site by MEMRI will still be possible several weeks post-lesioning. Additionally, in untreated controls, the signal intensity of the Mn^{2+}-enhanced proximal ON stump can be expected to be visibly diminished by 21dpl when compared to the acute stage after ONC (Fig. 1). With regard to treated, regenerating animals, visual inspection of the T1-weighted images alone may not be sufficient for identification of RGC axon-regenerative responses to treatment, especially distal to the lesion site as the density of regenerating RGC axons in the distal stump may be below detection threshold. However, CNR analysis of the MEMRI data along the ON can be expected to reveal axon-regenerative responses to treatment in terms of elevated CNR profiles proximal and/or distal to the ONC injury site.

3.2.5 Ivit Injection of Mn^{2+}-Alginate Microbeads

Ivit injection of Mn^{2+}-alginate microbeads (bead diameter ~500 μm, *see* **Note 6**) can be performed using the same protocol as for *ivit* $MnCl_2$ injections, using a Hamilton syringe with a 25G needle or a glass micropipette connected to a sterile 50 ml syringe by polyethylene tubing.

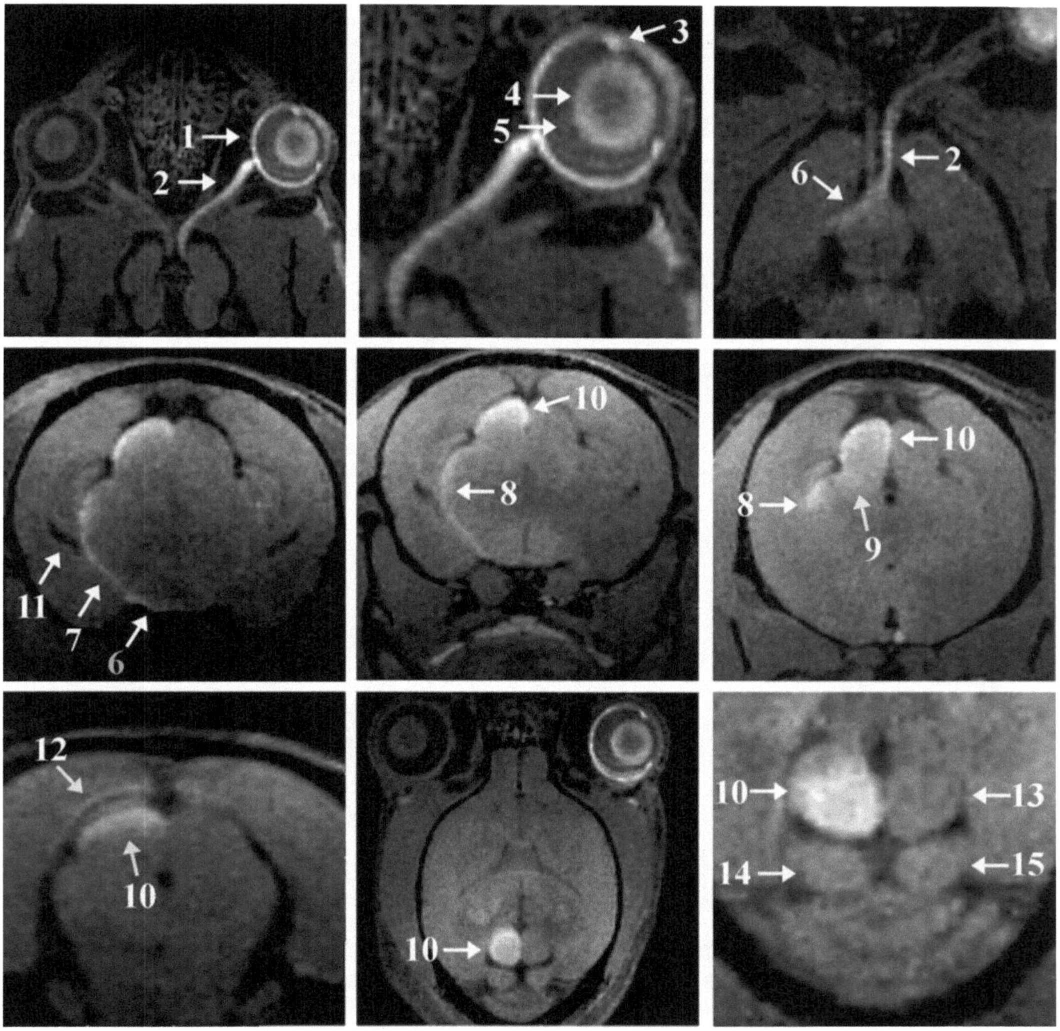

Fig. 2 MEMRI of the intact rat visual projection 24 h after *ivit* $MnCl_2$ injection showing Mn^{2+}-induced enhancement of the retina (1), ON (2), ciliary body (3), lens (4), vitreous body (5), optic chiasm (6) and contralateral optic tract (7), LGN (8), pretectal area (9), superior colliculus (10), and cortex (12). Some of the optic tract axons synapse in the lateral geniculate nucleus and postsynaptic axons project in the optic radiation (11). No Mn^{2+}-induced signal enhancement is observed in the contralateral inferior colliculus (14) or ipsilateral colliculi (13, 14). Reprinted from Thuen et al. [5], copyright 2005, with permission from Wiley

Expected Results

Successful Mn^{2+}-alginate microbead injection should not compromise the integrity of the eye. Clear Mn^{2+}-induced enhancement of the entire visual projection can be expected in intact animals 24–48 h after Mn^{2+}-alginate microbead injection (Fig. 3). The efficacy of such controlled-release systems in reducing *ivit* Mn^{2+} concentration compared to bolus Mn^{2+} injections is contingent on parameters such as the alginate type used, choice of cross-linking ions for Mn^{2+}-alginate microbead formation, concentration of bound Mn^{2+}

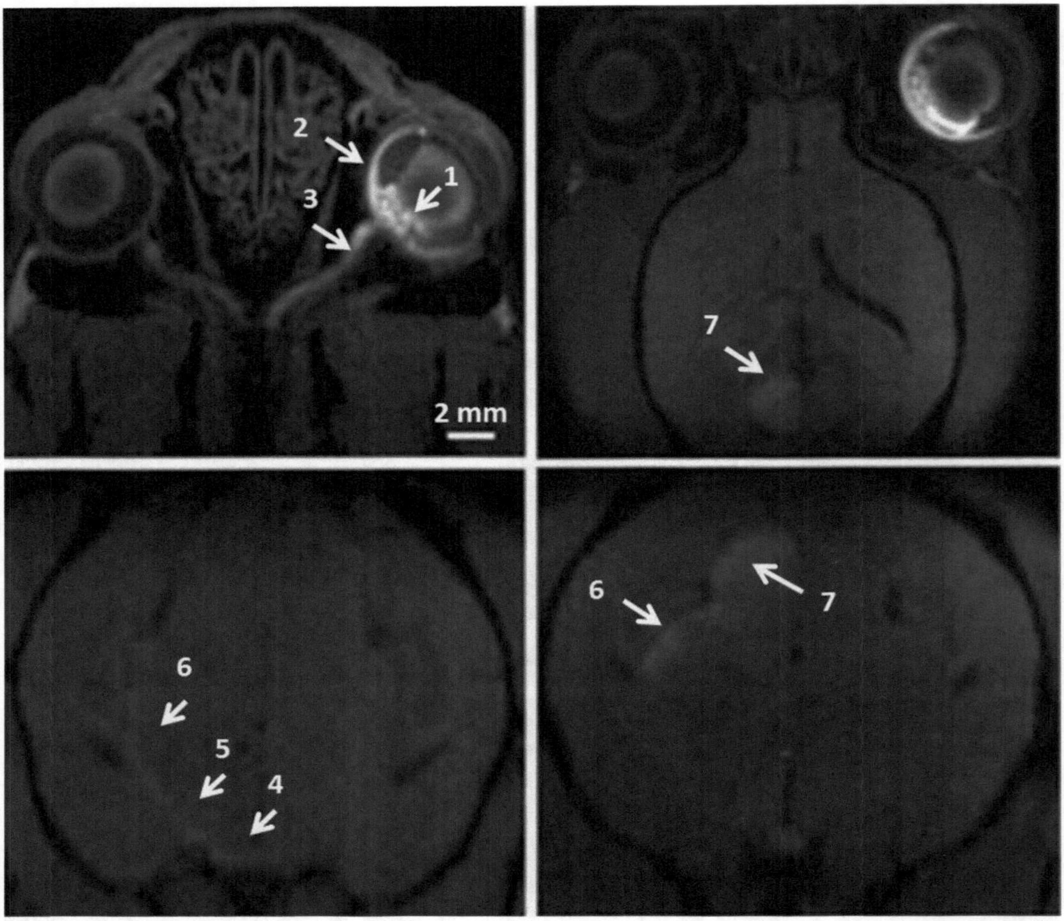

Fig. 3 MEMRI of the rat visual pathway after *ivit* injection of Mn^{2+}-alginate microbeads (1). Mn^{2+}-induced enhancement can be observed in the retina (2), ON (3), optic chiasm (4), as well as contralateral optic tract (5), LGN (6), and superior colliculus (7). Reprinted from Mørch et al. [52], copyright 2012, with permission from Wiley

per microbead, and number of microbeads injected in the vitreous body. We have previously used compartmental analysis [51] and estimated that *ivit* injection of approximately 20 Mn^{2+}-alginate microbeads formed using 0.1 M $MnCl_2$ in 1 mM $BaCl_2$ gelling solution (*see* **Note 6**) may achieve a reduction of *ivit* Mn^{2+} concentration of 44–63 % compared to 100 % of bolus injection [51]. The specific study constitutes proof of concept for the utility of Mn^{2+}-alginate microbeads as systems for slow release of Mn^{2+} for the purpose of optimizing contrast enhancement while minimizing the risk of Mn^{2+}-induced toxicity [51]. However, further refinement is required before the method can be used as an alternative to bolus *ivit* $MnCl_2$ injections for serial application in MEMRI of the regenerating ON.

4 Notes

1. Tissue paper bits facilitate stabilization of the ON for ONC injury. Inserts can easily be made by rolling/twisting small pieces of lab tissue paper (e.g., Kimwipes) between the thumb and index finger.
2. For *ivit* injections, either a pulled glass micropipette or a Hamilton syringe with a 25G needle can be used. The advantage of using a pulled glass micropipette is the ability to monitor advancement of the meniscus within the glass. Furthermore, the glass micropipette can be held between the thumb and index finger, allowing steady grip throughout the injection. A 10 μl pipette and fine long pipette tips can be used for backfilling of the glass micropipette. The micropipette can be connected to a 50 ml syringe by polyethylene tubing and the injection controlled by one hand, while the glass micropipette is held by the other. Alternatively, a Hamilton syringe can be used with a 25G needle directly attached to it or attached via polyethylene tubing. The latter allows monitoring of the meniscus advancing through the tubing using the Hamilton syringe scale.
3. As an alternative to fentanyl-fluanisone/midazolam, anaesthesia can be maintained with 2 % isoflurane during *ivit* injections. The use of fentanyl-fluanisone/midazolam enables full control of the animal on the rat head frame, including tilting of the head to allow better access to the eye for injection. Depending on the type of rat head frame used, the above may not always be possible as the isoflurane mask and adjacent tubing may impede access to the eye. As an alternative, injections under isoflurane anaesthesia may be made without fixing the animal to a head frame; however, the frame enables better control and stability during the procedure.
4. The neuroanatomical features of the RGC axonal projection suggest that the relative depth of the ON vis-à-vis the surface coil will increase along the pathway. As a result, gradual loss of RF signal will occur. To correct for the gradual reduction of the RF signal along the ON which can be detected by the surface coil, additional T1-weighted 3D FLASH sequence scans are performed using coupled- and single-coil operations, applying the same sequence parameters as above, but with reduced resolution and acquisition time [6].
5. The technique involves binarizing the 3D image by a global threshold based on the mean signal of a region of interest (ROI) placed in the non-enhanced ON. The ON is subsequently segmented from the brain by six connected regions growing from a manually placed seed point within the Mn^{2+}-enhanced ON. A sliding box technique is used for identification of ON

coordinates, ensuring correct definitions, also at start and end points. 2D planes are re-sampled every 0.2 mm perpendicular to the ON and the Mn^{2+}-enhanced signal (S_{mn}) in each plane calculated. CNR are calculated from the formula CNR = 0.655 $(S_{mn} - S_0)/SD_{air}$, where S_0 represents the signal intensities in an ROI in the non-enhanced ON, and SD_{air} is the mean value of the SD in two ROI in air.

6. Alginates are naturally occurring polysaccharides consisting of unbranched copolymers of 1→4 linked β-d-mannuronic acid (M) and α-l-guluronic acid (G) [53]. Structural and compositional modification can be achieved through the conversion of M-residues to G-residues with the use of C-5 epimerases [54–58]. Mn^{2+}-alginate microbeads can be made by pumping a Na^{2+}-alginate solution through an electrostatic bead generator [51] and dripping it into a gelling solution containing 0.1 M $MnCl_2$, plus 1 mM $BaCl_2$ or 10 mM $CaCl_2$ [51]. The choice of Ca^{2+} or Ba^{2+} as cross-linking ions further determines microbead properties such as strength, stability, syneresis, and permeability [59]. As a result, different combinations of alginate type and cross-linking ions produce Mn^{2+}-alginate microbeads with different properties in terms of long-term stability and Mn^{2+} release profiles [51].

Acknowledgements

A number of people have made valuable contributions towards the protocols and techniques described in this chapter. The authors are grateful to Professor Martin Berry for expert training and advice in establishing the surgical protocols; Dr Christian Brekken, Dr Marte Thuen, and Professor Olav Haraldseth for their expertise in MRI; Dr Øystein Olsen for mathematical modeling; Dr Yrr Mørch and Professor Gudmund Skjåk-Bræk for the Mn^{2+}-alginate microbeads; and Mrs Tina Bugge Pedersen for technical assistance. This work was supported by the Norwegian Research Council, Centre for Research-Based Innovation, Medical Imaging Laboratory (MI Lab), NTNU, Norway, and Functional Genomics (FUGE), Norway.

References

1. Harvey AR, Hu Y, Leaver SG et al (2006) Gene therapy and transplantation in CNS repair: the visual system. Prog Retin Eye Res 25:449–489
2. Plant GW, Harvey AR, Leaver SG et al (2011) Olfactory ensheathing glia: repairing injury to the mammalian visual system. Exp Neurol 229:99–108
3. Berry M, Ahmed Z, Lorber B et al (2008) Regeneration of axons in the visual system. Restor Neurol Neurosci 26:147–174
4. Sandvig A, Sandvig I, Berry M et al (2011) Axonal tracing of the normal and regenerating visual pathway of mouse, rat, frog, and fish using manganese-enhanced magnetic resonance

imaging (MEMRI). J Magn Reson Imaging 34: 670–675

5. Thuen M, Singstad TE, Pedersen TB et al (2005) Manganese-enhanced MRI of the optic visual pathway and optic nerve injury in adult rats. J Magn Reson Imaging 4:492–500
6. Sandvig I, Thuen M, Hoang L et al (2012) In vivo MRI of olfactory ensheathing cell grafts and regenerating axons in transplant mediated repair of the adult rat optic nerve. NMR Biomed 26:620–631
7. Bear MF, Connors BW, Paradiso MA (2006) The central nervous system. In: Neuroscience. Exploring the brain. 3rd. rev. ed. Lippincott Williams & Wilkjins, Baltimore
8. Lorber B, Berry M, Logan A (2008) Different factors promote regeneration of adult rat retinal ganglion cells after lens injury and intravitreal peripheral nerve grafting. J Neurosci Res 86:894–903
9. Zhi Y, Lu Q, Zhang CW et al (2005) Different optic nerve injury sites result in different responses of retinal ganglion cells to brain-derived neurotrophic factor but not neurotrophin-4/5. Brain Res 1047:224–232
10. Zeng BY, Anderson PN, Campbell G et al (1995) Regenerative and other responses to injury in the retinal stump of the optic nerve in adult albino rats: transection of the intracranial optic nerve. J Anat 189:495–508
11. Cho EY, So K-F (1993) Sprouting of axon-like processes from axotomised retinal ganglion cells is influenced by the distance of axotomy from the cell body and the mode of transplantation of the peripheral nerve. Restor Neurol Neurosci 6:29–34
12. Bähr M (2000) Live or let die—retinal ganglion cell death and survival during development and in the lesioned adult CNS. Trends Neurosci 23:483–490
13. Isenmann S, Wahl C, Krajewski S et al (1997) Upregulation of Bax protein in degenerating retinal ganglion cells precedes apoptotic death after optic nerve lesion in the rat. Eur J Neurosci 9:1763–1772
14. Berry M (1982) Post-injury breakdown products inhibit axonal growth: an hypothesis to explain the failure of axonal regeneration in the mammalian central nervous system. Bibl Anat 23:1–11
15. Sofroniew MV (2009) Molecular dissection of reactive astrogliosis and glial scar formation. Trends Neurosci 32:638–647
16. Fawcett JW (2006) The glial response to injury and its role in the inhibition of CNS repair. Adv Exp Med Biol 557:11–24
17. Rhodes KE, Fawcett JW (2004) Chondroitin sulphate proteoglycans: preventing plasticity or protecting the CNS? J Anat 204:33–48
18. Benowitz LI, Yin Y (2007) Combinatorial treatments for promoting axon regeneration in the CNS: strategies for overcoming inhibitory signals and activating neurons' intrinsic growth state. Dev Neurobiol 67:1148–1165
19. Logan A, Ahmed Z, Baird A et al (2006) Neurotrophic factor synergy is required for neuronal survival and disinhibited axon regeneration after CNS injury. Brain 129: 490–502
20. Berry M, Carlile A, Hunter A et al (1999) Optic nerve regeneration after intravitreal peripheral nerve implants: trajectories of axons regrowing through the optic chiasm into the optic tracts. J Neurocytol 28:721–741
21. Berry M, Carlile J, Hunter A (1996) Peripheral nerve explants grafted into the vitreous body of the eye promote the regeneration of retinal ganglion cell axons severed in the optic nerve. J Neurocytol 25:147–170
22. So KF, Aguayo AJ (1985) Lengthy regrowth of cut axons from ganglion cells from peripheral nerve transplantation into the retina of adult rats. Brain Res 328:349–354
23. Aguayo AJ, David S, Bray GM (1981) Influences of the glial environment on the elongation of axons after injury; transplantation studies in adult rodents. J Exp Biol 95: 231–240
24. David S, Aguayo AJ (1981) Axonal elongation into peripheral nervous system 'bridges' after central nervous system injury in adult rats. Science 214:931–933
25. Richardson PM, McGuinness UM, Aguayo AJ (1980) Axons from CNS neurons regenerate into PNS grafts. Nature 284:264–265
26. Ahmed Z, Aslam M, Lorber B et al (2010) Optic nerve and vitreal inflammation are both RGC neuroprotective but only the latter is RGC axogenic. Neurobiol Dis 37:441–454
27. Lorber B, Berry M, Logan A (2005) Lens injury stimulates adult mouse retinal ganglion cell axon regeneration via both macrophage- and lens-derived factors. Eur J Neurosci 21: 2029–2034
28. Fischer D, Heiduschka P, Thanos S (2001) Lens-injury stimulated axonal regeneration throughout the optic pathway of adult rats. Exp Neurol 172:257–272
29. Leon S, Yin Y, Nguyen J et al (2000) Lens injury stimulates axon regeneration in the mature rat optic nerve. J Neurosci 20:4615–4626
30. Fischer D, Pavlidis M, Thanos S (2000) Cataractogenic lens injury prevents traumatic

ganglion cell death and promotes axonal regeneration both in vivo and in culture. Invest Ophthalmol Vis Sci 41:3943–3954

31. Park KK, Liu K, Hu Y et al (2010) PTEN/mTOR and axon regeneration. Exp Neurol 223:45–50

32. Kurimoto T, Yin Y, Omura K et al (2010) Long distance axon regeneration in the mature optic nerve: contributions of oncomodulin, cAMP, and *pten* gene deletion. J Neurosci 30: 15654–15663

33. Li Y, Sauvé Y, Li D et al (2003) Transplanted olfactory ensheathing cells promote regeneration of cut adult rat optic nerve axons. J Neurosci 23:7783–7788

34. Li Y, Li D, Raisman G (2007) Transplanted Schwann cells, not olfactory ensheathing cells myelinate optic nerve axons. Glia 55:312–316

35. Wu MM, Fan DG, Tadmori I et al (2010) Death of axotomised retinal ganglion cells delayed after intraoptic nerve transplantation of olfactory ensheathing cells in adult rats. Cell Transplant 19:159–166

36. Liu Y, Gong Z, Liu L et al (2010) Combined effect of olfactory ensheathing cell (OEC) transplantation and glial cell line-derived neurotrophic factor (GDNF) intravitreal injection on optic nerve injury in rats. Mol Vis 16: 2903–2910

37. Thuen M, Olsen O, Berry M et al (2009) Combination of Mn(2+)-enhanced and diffusion tensor MR imaging gives complementary information about injury and regeneration in the adult rat optic nerve. J Magn Reson Imaging 1:39–51

38. Thuen M, Berry M, Pedersen TB et al (2008) Manganese-enhanced MRI of the rat visual pathway: acute neural toxicity, contrast enhancement, axon resolution, axonal transport, and clearance of Mn(2+). J Magn Reson Imaging 4:855–865

39. Sloot WN, Gramsbergen JP (1994) Axonal transport of manganese and its relevance to selective neurotoxicity in the rat basal ganglia. Brain Res 657:124–132

40. Pautler RG, Silva AC, Koretsky AP (1998) In vivo neuronal tract-tracing using manganese-enhanced magnetic resonance imaging. Magn Reson Med 40:740–748

41. Narita K, Kawasaki F, Kita H (1990) Mn and Mg influxes through Ca channels of motor nerve terminals are prevented by verapamil in frogs. Brain Res 510:289–295

42. Finley JW (1998) Manganese uptake and release by cultured human hepato-carcinoma (hep-G2) cells. Biol Trace Elem Res 64: 101–118

43. Olsen Ø, Kristoffersen A, Thuen M et al (2010) Manganese transport in the rat optic nerve evaluated with spatial and time-resolved magnetic resonance imaging. J Magn Reson Imaging 32:551–560

44. Goldstein LS, Yang Z (2000) Microtubule-based transport systems in neurons: the roles of kinesins and dyneins. Ann Rev Neurosci 23: 39–71

45. Pautler RG, Mongeau R, Jacobs RE (2003) In vivo trans-synaptic tract-tracing from the murine striatum and amygdala utilizing manganese-enhanced MRI (MEMRI). Magn Reson Med 50:33–39

46. Saleem KS, Pauls JM, Augath M et al (2002) Magnetic resonance imaging of neuronal connections in the macaque monkey. Neurotechnique 34:685–700

47. Watanabe T, Michaelis T, Frahm J (2001) Mapping retinal projections in the living rat using high resolution 3D gradient-echo MRI with Mn^{2+}-induced contrast. Magn Reson Med 46:424–429

48. Liang XY, Cheung SW, Chan KC et al (2011) CNS regeneration after chronic injury using a self-assembled nanomaterial and MEMRI for real-time in vivo monitoring. Nanomedicine 7:351–359

49. Chan KC, Fu QL, Hui ES et al (2008) Evaluation of the retina and optic-nerve in a rat model of chronic glaucoma using in vivo manganese-enhanced magnetic resonance imaging. Neuroimage 40:1166–1174

50. Santamaria AB (2008) Manganese exposure, essentiality and toxicity. Indian J Med Res 128: 484–500

51. Mørch YA, Sandvig I, Olsen O et al (2012) Mn-alginate gels as a novel system for controlled release of Mn2+ in manganese-enhanced MRI. Contrast Media Mol Imaging 7:265–275

52. Olsen Ø, Thuen M, Berry M et al (2008) Axon tracing in the adult rat optic nerve and tract after intravitreal injection of MnDPDP using a semiautomatic segmentation technique. J Magn Reson Imaging 27:34–42

53. Haug A, Larsen B, Smidsrød O (1996) A study of the construction of alginic acid by partial hydrolysis. Acta Chem Scand 20:183–190

54. Mørch YA, Holtan S, Donati I et al (2008) Mechanical properties of C-5 epimerised alginates. Biomacromolecules 9:2360–2368

55. Mørch YA, Donati I, Strand BL et al (2005) Molecular engineering as an approach to design new functional properties of alginate. Biomacromolecules 8:2809–2814
56. Donati I, Holtan S, Mørch YA et al (2005) New hypothesis on the role of alternating sequences in calcium-alginate gels. Biomacromolecules 6:1031–1040
57. Strand BL, Mørch YA, Syvertsen KR et al (2003) Microcapsules made by enzymatically tailored alginate. J Biomed Mater Res A 64: 540–550
58. King A, Strand BL, Rokstad AM et al (2003) Improvement of the biocompatibility of alginate/poly-L-lysine/alginate microcapsules by the use of epimerised alginate as a coating. J Biomed Mater Res A 64:533–539
59. Mørch YA, Donati I, Strand BL et al (2006) Effect of Ca2+, Ba2+, and Sr2+ on alginate microbeads. Biomacromolecules 7: 1471–1480

Index

Andrew J. Murray (ed.), *Axon Growth and Regeneration: Methods and Protocols*, Methods in Molecular Biology, vol. 1162, DOI 10.1007/978-1-4939-0777-9, © Springer Science+Business Media New York 2014

S

T

V

W

MIX
Papier aus verantwortungsvollen Quellen
Paper from responsible sources
FSC® C105338

If you have any concerns about our products,
you can contact us on
ProductSafety@springernature.com

In case Publisher is established outside the EU,
the EU authorized representative is:
Springer Nature Customer Service Center GmbH
Europaplatz 3, 69115 Heidelberg, Germany

Printed by Libri Plureos GmbH
in Hamburg, Germany